ARSENIC

RISKS OF EXPOSURE, BEHAVIOR IN THE ENVIRONMENT AND TOXICOLOGY

CHEMISTRY RESEARCH AND APPLICATIONS

Additional books in this series can be found on Nova's website
under the Series tab.

Additional e-books in this series can be found on Nova's website
under the eBooks tab.

CHEMISTRY RESEARCH AND APPLICATIONS

ARSENIC

RISKS OF EXPOSURE, BEHAVIOR IN THE ENVIRONMENT AND TOXICOLOGY

RATKO KNEŽEVIĆ
EDITOR

science publishers
New York

Library of Congress Cataloging-in-Publication Data

ISBN: 978-1-53612-461-3

Published by Nova Science Publishers, Inc. † New York

CONTENTS

PREFACE

Arsenic (As) is one of the most toxic elements for plants, animal and humans, found in waters. It is considered as carcinogenic and causative agent of numerous human diseases by the International Agency for Research on Cancer (IARC, group 1) since 1980. Based on these observations and evidences, many studies have highlighted the interest of using plants for the detection of arsenic in the environment. Arsenic sources and speciation, as well as arsenic uptake and toxicity in plants, will be reviewed in Chapter One. Chapter Two aims to introduce the arsenic uptake in some of the common and popular fruits and vegetables that have been a pathway of risk to human health. In Chapter Three a three-module continuous plant design, capable of deliver up to 1 m3/day of arsenic-free drinking water, is described. Chapter Four studies the toxicogenesis of arsenic in wheat and paddy plants with subsequent internal metabolism and remediation strategies.

Chapter 1 - Arsenic (As) is one of the most toxic elements for plants, animals and humans. It has been considered as carcinogenic and causative agent of numerous human diseases by the International Agency for Research on Cancer since 1980 (IARC, 2012). Depending on the reaction with other molecules present in air, soil or water, arsenic exists in the environment in a variety of species including inorganic forms [arsenate (AsO_4^{3-})/arsenite (AsO_3^{3-})] and organic forms (methylated anionic species and organo-arsenic species in food). This speciation gives specific toxicity and uptake rate of this pollutant by organisms. Considering only plants, inorganic arsenic is easily taken up by plant roots as arsenate is chemically similar to phosphates and competes for plant phosphate transporters at the cell membrane interface. With arsenite, it has been suggested that it is taken up by aquaglyceroporins which are also responsible for the uptake of other arsenic compounds, such as monomethylarsonic acid (MMA) and dimethylarsinic acid (DMA). Once in plants, arsenic is accumulated mainly in the root system, but it is also transferred in the upper parts of plants. Indeed, arsenic accumulation in plants causes important physiological damages to the chlorophyllian parts since this toxic compound is well known to destroy the chloroplast membrane, to reduce chlorophyll content and so to affect photosynthesis. Moreover, arsenic exposure reduces osmotic potential and generally induces an oxidative stress leading to the production of reactive oxygen species (ROS). Apart from these dramatic changes for the plant development, arsenic can also cause genotoxic effects contributing to micronuclei formation and chromosomal aberrations.

Based on these observations and evidences, many studies have highlighted the interest of using plants for the detection of arsenic in the environment. Therefore, arsenic sources and speciation as well as arsenic uptake and toxicity in plants will be reviewed in this chapter. The plant tolerance and defense mechanisms to arsenic will be also investigated with a special focus on the aquatic environment.

Chapter 2 - Arsenic is a food chain contaminant. Arsenic contamination of fruits and vegetables is now being the additional source of this toxic element to mankind in South and South-east Asia. Current evidence shows that arsenic uptake in dietary plants is proportionally related to the presence of arsenic in soils and irrigation water where the major factors are flooding, arsenic forms, microorganism, organic matter, origin, and type of soil and plant. Excessive accumulation of arsenic in fruits and vegetables poses a potential health risk to the populations with higher consumption of fruits and vegetables, particularly to the local population of arsenic-affected area. To date, little attention has been paid to the risk of using arsenic-contaminated fruits and vegetables. In this perspective, the authors aim is to introduce the arsenic uptake in some of the common and popular fruits and vegetables that have been a pathway of risk to human health.

Chapter 3 - Arsenic (As) strongly limits water potability due to its high toxicity. The Chaco-Pampean plain is one of the regions worldwide recognized for its high arsenic content in groundwater and the involved area covers about 106 km^2 in Argentina. Arsenic levels in groundwater above 100 μg/L have been frequently reported. For this reason, it is imperative to develop efficient and inexpensive technical solutions for the elimination of arsenic from drinking water.

In the present work a three-module continuous plant design, capable of deliver up to 1 m^3/day of arsenic-free drinking water, is described. The system, whose first and main stage is a Zero-Valent Iron (ZVI) reactive bed, is simple, easy to use, was designed to respond small communities' needs and can be adapted for groundwater with different physicochemical characteristics.

Arsenic removal by ZVI-based technologies is related to the corrosion products generated by metallic iron oxidation and involves different mechanisms, including adsorption, surface complexation, surface precipitation and co-precipitation. Iron corrosion rates depend on both the operating conditions and the ZVI source used. Therefore, the effects of changing the main operational variables, of columns packed with iron wool, were analyzed in order to select the most favorable settings.

In many cases, the application of ZVI-based techniques is limited by reactivity losses and reductions of hydraulic conductivity caused by the accumulation of corrosion products. These problems arise due to the formation a thick layer of iron oxides onto ZVI surface, especially in natural waters with relatively high dissolved oxygen content. Consequently, the hydraulic behavior of the designed plant was studied throughout the operation period. In addition, stimulus response tests were carried out periodically to determine the residence time distribution along the reactive column.

The results obtained show that the lifespan of the plant may be predicted by taking in to account both the main chemical processes involved and the fluid dynamic properties of the bed.

Chapter 4 - The metalloid arsenic (As) is a natural constituent, contaminating the agro-ecosystem and gets increased along with the expansion of human health risks due to anthropogenic pollution. Rice and wheat are the staple cereal crops that are being cultivated in arsenic contaminated fields globally. From the very seedling stage to the final harvesting of crops, the effect of arsenic is prominently resulting in declination of root-shoot length and enzymatic expressivity in both wheat and paddy plants. Antioxidant enzymatic responses viz. superoxide dismutase (SOD), ascorbate peroxidase (APX), glutathione reductase (GR), and a few, get compromised immensely due to the excess bioavailability of arsenic to the plant system. Speciation of arsenic in its As(III) and As(V) forms are predominant respectively in aqueous and aerobic soil profile entangling with other elements in the system for the transportation inside the plant. Plants like rice and wheat tend to accumulate a greater percentage of arsenite [As(III)] due to its higher mobility and solubility that let the access of transportation through plant transporter molecules known as nodulin-like intrinsic proteins (NIPs) or aquaporins and Lsi transporters. Arsenate [AS(V)] commonly compete with phosphate molecule for the passage through Pht molecules by having a similar molecular structure. Apart from the inorganic species, organic forms of arsenic might get passed through the plant uptake system but less effectively. Though wheat and paddy plants are not hyperaccumulators of arsenic, some of the wetland plants namely, *Pteris* spp., *Eichhornia* spp., Agrostis spp., can be helpful in course of phytoremediation of arsenic before rice or wheat cultivation on contaminated fields. Phytochelatins are plant derived compounds, a derivative of glutathione that might form chelates with arsenic and can be a mode of plant detoxification of excess arsenic accumulation inside the plant tissue system. This glimpse of previous literature and current studies will enlighten the toxicogenesis of arsenic in wheat and paddy plant with subsequent internal metabolism and remediation strategies furthermore.

In: Arsenic
Editor: Ratko Knežević

ISBN: 978-1-53612-461-3
© 2017 Nova Science Publishers, Inc.

Chapter 1

THE ARSENIC THREAT IN AQUATIC ENVIRONMENTS: THE PLANT'S POINT OF VIEW

Maha Krayem[1,2,*], Véronique Deluchat[1],
Raphaël Decou[1] and Pascal Labrousse[1,†]
[1]GRESE - EA 4330, Université de Limoges,
Limoges, France
[2]PRASE-EDST, Lebanese University,
Beirut, Lebanon

ABSTRACT

Arsenic (As) is one of the most toxic elements for plants, animals and humans. It has been considered as carcinogenic and causative agent of numerous human diseases by the International Agency for Research on Cancer since 1980 (IARC, 2012). Depending on the reaction with other molecules present in air, soil or water, arsenic exists in the environment in a variety of species including inorganic forms [arsenate

---
[*] Corresponding author: maha.krayem@hotmail.com.
[†] Corresponding author: pascal.labrousse@unilim.fr.

$(AsO_4{}^{3-})$/arsenite $(AsO_3{}^{3-})$] and organic forms (methylated anionic species and organo-arsenic species in food). This speciation gives specific toxicity and uptake rate of this pollutant by organisms. Considering only plants, inorganic arsenic is easily taken up by plant roots as arsenate is chemically similar to phosphates and competes for plant phosphate transporters at the cell membrane interface. With arsenite, it has been suggested that it is taken up by aquaglyceroporins which are also responsible for the uptake of other arsenic compounds, such as monomethylarsonic acid (MMA) and dimethylarsinic acid (DMA). Once in plants, arsenic is accumulated mainly in the root system, but it is also transferred in the upper parts of plants. Indeed, arsenic accumulation in plants causes important physiological damages to the chlorophyllian parts since this toxic compound is well known to destroy the chloroplast membrane, to reduce chlorophyll content and so to affect photosynthesis. Moreover, arsenic exposure reduces osmotic potential and generally induces an oxidative stress leading to the production of reactive oxygen species (ROS). Apart from these dramatic changes for the plant development, arsenic can also cause genotoxic effects contributing to micronuclei formation and chromosomal aberrations.

Based on these observations and evidences, many studies have highlighted the interest of using plants for the detection of arsenic in the environment. Therefore, arsenic sources and speciation as well as arsenic uptake and toxicity in plants will be reviewed in this chapter. The plant tolerance and defense mechanisms to arsenic will be also investigated with a special focus on the aquatic environment.

Keywords: arsenic, speciation, phytotoxicity, defense, biomarkers, aquatic macrophytes

1. Introduction

Water quality influences human health as well as aquatic life forms (fauna and flora), yet this water quality is endangered by many inorganic and organic pollutants through natural processes (erosion, evaporation, runoff of rainwater) and human activities (agricultural, industrial and domestic). These factors can cause physical, chemical and biological degradation of the aquatic systems and thus affecting the ecosystem services. In Europe, the Water Framework Directive (WFD) adopted by the

European Council and Parliament on the 23rd October 2000 established a framework for water management. This framework aimed to ensure both good ecological and chemical status for surface and ground waters in 2015. In addition, 33 hazardous substances listed by the WFD must be phased out by 2020 or must have a concentration in aquatic environments complying with the environmental quality standards (EQS) set out in this directive and daughter directives (2008/105/EC, 2013/39/EC). In order to maintain and improve water quality, it is imperative to develop tools enabling the detection of pollutants as early as possible before the ecosystem is degraded. Usual pollutant detection uses analytical techniques that allows measurement and quantification of contaminants, however they are limited by: (i) the limit of quantification that can be higher compared to the concentration of contaminant that can affect the ecosystem, (ii) the point sampling (geolocation and time) may not be representative of pollution levels in water, (iii) the cost of sampling is relatively high due to specific equipment and skilled labor needs and (iv) the problems of matrix effects related to sample composition. To avoid these problems, other methods based on chemical, physical and biological parameters of living organisms have been developed. They are called "biological indicators" or "bioindicators" and their changes constitute biomarkers of pollution. When affected by poor water quality, a bioindicator exhibits changes at either population or individual level (i.e., morphological, physiological and/or biochemical). These modifications may even appear at very low concentrations not detectable with chemical analysis. For some parameters like population-based indices or morphology, the monitoring is relatively easy to perform and cost effective.

Ecotoxicology is the science devoted to the study and prevention of the harmful effects of natural or anthropogenic chemicals on the structure, functions and biodiversity of ecosystems (Tarazona and Ramos-Peralonso, 2014). Within this sphere, biological indexing is used to detect and evaluate the toxic effects of pollutants. It began with fish, insect larvae and diatoms that showed they were resistant and tolerant to pollution thus allowing biological indexing through population analysis (River Fish

Index, Standardized Global Biological Index and Diatom Biotic Index). Recently, aquatic macrophytes have been integrated as bioindicator in the EU Water Framework Directive (Aguiar et al., 2014). Among these macrophytes, *Myriophyllum alterniflorum* presents morphological and physiological modifications when exposed to xenobiotic stresses by copper, arsenic and cadmium (Delmail, 2011; Krayem, 2015) and trophic modification (Chatenet et al., 2006; Krayem, 2015). This chapter is devoted to establish a state of art on arsenic pollution in freshwater, including its speciation in surface waters, arsenic-plant interactions (absorption, translocation…) and its effects on plant biomarkers.

2. POLLUTANTS IN THE AQUATIC ENVIRONMENT

Every year, thousands of tons of pollutants are discharged into the environment. The major contaminants present are polycyclic aromatic hydrocarbons (PAHs), polychlorinated biphenyls (PCBs), drug molecules, pesticides and metals (INERIS, 2014; RNO, 2006).

Figure 1. Sources of pollutants in the environment (adapted from Champeau, 2005).

These pollutants found in the earth's crust or in the atmosphere originate from natural or anthropogenic processes like industries (paints, printing, plastics, tanneries...), mining activities, automobile traffic, agriculture, domestic, radioactive waste treatment, etc. Large amount of xenobiotics (i.e., are defined as foreign substances to biological systems) are also transferred to the aquatic environment through soil leaching. Xenobiotic absorption is toxic and leads to non-tolerant species extinction. Conversely, tolerant species become dominant as they are able to cope with pollutant (Pollard et al., 2014; Van der Ent et al., 2013). In addition to biodiversity reduction, xenobiotics disturb biogeochemical cycles and contaminate food webs thus constituting a threat to human health (Liehr et al., 2005; Figure 1).

2.1. Trace Metals and Metalloids

Trace metals correspond to metals with lower concentration than 1 mg.kg^{-1} soil (Duffus, 2002). Metals include chemical elements found in rows IA, IIA and in lines of transition elements of the periodic table. Metalloids refer to chemical elements presenting intermediate physicochemical characteristics between metals and nonmetals (Kelter et al., 2008; Zangi and Filella, 2012). Seven elements are recognized as metalloids: Boron (B), Silicon (Si), Germanium (Ge), Arsenic (As), Antimony (Sb), Tellurium (Te) and Astatine (At). Naturally, trace metals and metalloids are present in rocks and ores of the earth's crust, usually as oxides, carbonates, silicates or sulphides. In a watershed, rock alteration and erosion lead to the presence of metals in the aquatic environment known as "geochemical background," i.e., concentration of natural contaminants in the surrounding environment. Volcanism, wildfires and hot springs also contribute to metals and metalloids dissemination (Baize and Sterckeman, 2001; Sanità di Toppi and Gabbrielli, 1999). In addition to these natural sources of contamination, anthropic activities contribute to increase metal and metalloid concentrations in the environment. As they are not decomposed by organisms, trace metals and metalloids are persistent in the environment (Daby, 2006). Two classes of metals are distinguished (Figure 2). Firstly, essential metals known as trace elements are necessary at low doses to support metabolic activities through metal-bearing enzymes for example. However, at higher doses, these essential elements present acute or chronic toxicity. Secondly, non-essential metals do not contribute to biological function and present acute or chronic toxicity according to dose exposure (Daby, 2006; Filipiak-Szok et al., 2015; Wood, 2011). Arsenic (As), cobalt (Co), cadmium (Cd), chromium (Cr) and nickel (Ni) can also develop carcinogenic properties (Chen and White, 2004; Hartwig, 1998). However, evidence shows that cadmium, as a non-essential metal, is involved in diatom metabolism. Indeed, the marine diatom, *Thalassiosira weissflogii,* uses cadmium in carbonic anhydrase, a key enzyme for its growth (Lane et al., 2005).

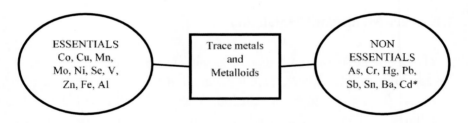

Figure 2. Subdivision of trace metallic and metalloid elements according to their nutritional importance in living organisms (Filipiak-Szok et al., 2015; Raskin et al., 1994).

Regarding more precisely the fate of these toxic compounds, trace metals and metalloids are both absorbed by plants and animals and later eliminated via excretion, but they can build up in tissues and then throughout the food chain. Thus, they represent a threat to human health (Liehr et al., 2005). In aquatic organisms, metal accumulation induces malformations, tissue necrosis, organ ulceration and atrophy, perturbations of cellular metabolism, photosynthesis and DNA repair systems (Calow, 2009; Clemens and Ma, 2016). The degree of toxicity depends on several parameters such as (i) the mechanism of plant absorption, (ii) the characteristics of the studied organism (species, sex, age, physiological development stage, etc.) and (iii) the metal concentration in a given organ

(Amiard-Triquet and Rainbow, 2009; Calow, 2009; Rand, 1995). For this reason, knowledge of the quantity of metals within organisms and ecosystems is of major interest and a main goal of ecotoxicological studies (Huang et al., 2006). In the environment and more particularly in the aquatic ecosystem, numerous trace metals and metalloids like manganese (Mn), boron (B), zinc (Zn), mercury (Hg), copper (Cu), chromium (Cr), nickel (Ni), aluminum (Al), lead (Pb), cadmium (Cd) and arsenic (As) are known for their toxicity (Dazy et al., 2009; Hawa Bibi et al., 2010; Kaiser, 2001; Rau et al., 2007). For example, the toxic effects of boron on the aquatic Araceae *Lemna minor* and *Lemna gibba* were assessed and showed that this metal reduced the growth of *Lemna minor* after an exposure to 2.3 mM for 7 days (Böcük et al., 2013). This was confirmed by similar results showing that an exposure of 7 days to 0.75 mM of B affected not only the growth of *Lemna minor* and *Lemna gibba* but also activated the antioxidative system in these plants (Gür et al. 2016). Similarly, Cr is very well studied for its toxic effects on aquatic plants as observed in the aquatic moss *Fontinalis antipyretica* (Dazy et al., 2008) and in aquatic macrophytes *Alternanthera philoxeroides, Borreria scabiosoides, Polygonum ferrugineum, Eichhornia crassipes* (Mangabeira et al., 2011) and *Callitriche cophocarpa* (Augustynowicz et al., 2016). Due to the huge range of elements, their complex relationship with phytotoxicity and their uptake by plants, this review will focus specifically on As which is particularly well known for its toxicity for all organisms even at very low concentrations and sometimes, its high concentrations in waters (Hasanuzzaman et al., 2015).

2.2. Origin, Speciation and Biological Roles of Arsenic

As is a natural element (0.00021% in the earth's crust) appearing as a silvery crystalline solid in its pure form (Tangahu et al., 2011). It is a metalloid with atomic number 33 and molar mass of 74.9216 g.mol^{-1}. The quality standard for concentration in aquatic environment is 4 µg.L^{-1} (MEDD et Agences de l'eau, 2003). Each year, volcanism transfers naturally between 2 800 and 8 000 tons of arsenic in the environment. Other natural sources of arsenic are wildfires, geochemical background by alteration of parent rock and dispersion of primary arsenic minerals like arsenopyrite (FeAsS), realgar (AsS) and orpiment (As_2S_3) (Laperche et al., 2003; Punshon et al., 2017). The largest quantity of arsenic originates from anthropogenic activities like combustion of municipal solid wastes, fossil fuels, metal smelting and direct use of arsenic herbicides (Punshon et al., 2017; Tangahu et al., 2011). Arsenic occurs under four degrees of oxidation: -III, 0, +III and +V and the toxicity of arsenic compounds decrease as follows: AsH_3 gas>As (-III)> As (III)> As (V)> methylated compounds (Sharma and Sohn, 2009). In the environment, arsenic can be distinguished in: (i) organic forms, which are mainly arsenobetaine, arsenocholine and mono and dimethyl derivatives (monomethylarsinous acid, dimethylarsinous acid, monomethylarsonic acid and dimethylarsinic acid) and (ii) inorganic compounds constituted essentially of arsenite (AsIII) and arsenate (AsV) (Farooq et al., 2016; Singh et al., 2015; Tangahu et al., 2011).

Table 1. Principal As derivatives in the environment and their pKa (Michon et al., 2007)

Compounds	Molecular formula	Oxidation degree	pKa
Arsine (volatile form)	AsH_3	3-	
Monomethylarsine (MMA)	H_2CH_3As	3-	
Dimethylarsine (DMA)	$(CH_3)_2HAs$	3-	
Trimethylarsine (TMA)	$(CH_3)_3As$	3-	
Arsenous acid	H_3AsO_3	3+	9.29/12.1/13.4
Arsenic acid	H_3AsO_4	5+	2.24/6.96/11.5
Monomethylarsonic acid (MMAA)	$CH_3AsO(OH)_2$	5+	4.19/8.77
Dimethylarsinic Acid (DMAA)	$(CH_3)_2AsO(OH)$	5+	1.78/6.14
Trimethylarsine oxide (TMAO)	$(CH_3)_3AsOH$	5+	3.6
Arsenobetaine (AB)	$(CH_3)_3As(CH_2)COOH$	5+	4.7
Arsenocholin (AC)	$(CH_3)_3As(CH_2)_2OH$	5+	

The speciation and mobility of arsenic in water is related to pH and oxido-reduction potential. In natural waters, the main forms are arsenite under reducing conditions (groundwater) and arsenate in surface waters (Cullen and Reimer, 1989; Villaescusa and Bollinger, 2008). Organic forms may occasionally occur in surface waters. Unlike metals, arsenic in solution is essentially in anionic form; consequently, for As (V), $H_2AsO_4^-$ is dominant in oxidizing media with pH less than 6.9 whereas $HAsO_4^{2-}$ becomes dominant when the pH increases. $H_3AsO_4^0$ and AsO_4^{3-} are only present respectively in extremely acidic and alkaline media. For As (III), $H_3AsO_3^0$ predominates in reducing medium with pH less than 9.2 (Bousquet, 1997; Smedley and Kinniburgh, 2002; Figure 3).

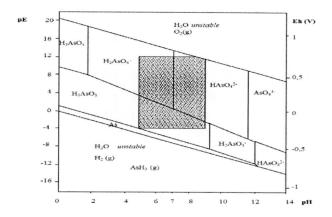

Figure 3. Eh-pH-pe speciation diagram of aqueous arsenic species at +25°C and 1 bar pressure (from Bousquet, 1997). (Eh: oxido-reduction potential).

It should be noted that not only chemical factors like pH, oxido-reduction potential (Eh), dissolved oxygen concentration, the presence of oxidizing and reducing agents, and organic matter content can influence the speciation of arsenic in the aquatic ecosystems. Other parameters like the presence of biological factors such as bacteria, phytoplankton and higher aquatic plants can participate in these chemical changes (Linnik, 2015). A significant intake of arsenic can be harmful to plants, animals and humans. According to the guidelines of USEPA (2001), the recommended limit of arsenic in drinking water is 10 $\mu g.L^{-1}$. Arsenic levels in surface and unpolluted groundwater generally vary between 1 and 10 $\mu g.L^{-1}$ (Singh et al., 2015). An exposure to 40 mg As.kg^{-1}soil is toxic for both terrestrial

Monocotyledons (corn, barley) and Dicotyledons (radish, cotton, peach, apple, apricot...) cultivated on sandy and loamy soils (Sheppard, 1992). Acute and chronic toxicity can occur after exposure to inorganic arsenic. In humans, symptoms of acute toxicity are gastro-intestinal disorders, nausea, vomiting, effects on the nervous system and skin. Chronic poisoning induces development of cancers targeting the skin, bladder, kidneys, lungs... Chronic intoxication with exposure to high levels of arsenic in drinking water induces gangrene like black foot disease (INERIS, 2010). Arsenic toxicity depends on its speciation and on the route of absorption. In humans, organic forms from food are not toxic as they are non-fixed by tissues and eliminated without transformation. Inorganic arsenic is much more toxic: (i) arsenite ion (AsO_3^3) due to its affinity with sulfhydryl groups (SH^-) in enzymes (Meharg and Hartley-Whitaker, 2002; Mishra et al., 2014); (ii) arsenate ion (AsO_4^{3-}) as it is analogous to phosphate ions, it can alter vital molecules and/or processes like ATP/ADP synthesis, DNA replication and enzymatic repair systems (Finnegan and Chen, 2012). Consequently, inorganic arsenic can induce indirectly genotoxic effects (Clemens and Ma, 2016; Sanders and Vermersch, 1982). Hence, As speciation has a great concern in ecotoxicological studies in aquatic organisms due to the differences observed in the stress level caused by the different species (Duester et al., 2011). Growth reduction and decline in fertility are symptoms of arsenic phytotoxicity following its absorption by plant roots (Garg and Singla, 2011). Perturbation of critical metabolic processes occur at high As doses as for example the As swap for phosphorus (P) in reactions of oxidative phosphorylations which can lead to plant death (Finnegan and Chen, 2012).

3. PLANTS/ARSENIC INTERACTIONS

3.1. Bioaccessibility of Arsenic to Plants

Metal bioaccessibility depends on both (i) its physicochemical speciation in the surrounding environment and (ii) its ability to diffuse towards the organism (limited by adsorption on other supports) and to be adsorbed on its surface. In aquatic environment, trace metals and metalloids are allocated between dissolved, colloidal and particulate phases. In this environment, they can be hydrated by water molecules, complexed with mineral or organic ligands and/or adsorbed on suspended macromolecules (Ardestani et al., 2014; Flemming and Trevors, 1989; Santore et al., 2001). All these reactions depend on several physicochemical factors like pH, temperature, ionic strength, oxidation state of metal, competitor ions and concentration of mineral and organic elements in the medium. For plant bioavailability and absorption, metals and metalloids must be present in free forms. Biotic ligand Models (BLM) allows determination of metal speciation and prediction of toxicity in aquatic medium of known composition (Di Toro et al., 2001; Santore et al., 2001; Smith et al., 2015; Figure 4).

3.2. Diffusion and Adsorption of Arsenic on Plant Surface

In aquatic ecosystems, a balanced distribution of trace metals and metalloids is established between water and plants. Thus, according to Fick's laws, diffusion flux is proportional to concentration gradient and molecular diffusion coefficient depending themselves on medium properties (temperature, viscosity, osmolarity, etc.) and distance. In addition, root exudates containing organic acids (phenolic-, malic-, tartaric- and succinic acids), terpenes, esters and/or amino acids (proline, threonine, aspartic acid, glutamic acid) play a major role in metal bioavailability (Koo et al., 2010; Xu et al., 2007). Indeed, the exudate composition can modify speciation, adsorption, desorption and trace metal migration in the rhizosphere by changing the chemical properties of the compounds, and thus their bioavailability for adsorption then absorption (Xu et al., 2007). Adsorption, the first "touch" between chemical compounds and plant surface, is a surface phenomenon involving multiple interactions. Weak or strong bonds can be established like hydrogen bonds, Van Der Waals interactions or covalent bonding through functional groups

(-OH, -NH$_2$ and -SH) located on the cell wall (Rand, 1995). Trace metals and metalloids can also be adsorbed or precipitated at the root interface by iron oxi-hydroxides constituting the well-known iron plaque. This is mainly involved in As adsorption, thus preventing or enhancing the As uptake by plant roots (Guan et al., 2009; Punshon et al., 2017). Other anions like carbonates, sulfates, phosphates, silicates as well as organic matter can also affect this adsorption through competition mechanisms (Meng et al., 2000).

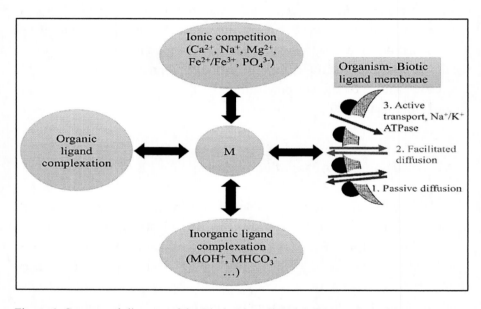

Figure 4. Conceptual diagram of the Biotic Ligand Model (BLM; adapted from Di Toro et al., 2001; Smith et al., 2015). ATPase: Adenosine TriPhosphatase, M: metal/metalloid.

3.3. Arsenic Assimilation by Plants

Chemical elements enter the plant through a process known as assimilation or absorption. Compared to adsorption, absorption is slower and sometimes requires energy as chemical compounds must pass through plant biological barriers: the plasma membrane and the cell wall.

Apoplastic molecules can reach the symplast via three independent main processes (Figure 4):

1. Passive diffusion is an "energyless" transport, dependent on the concentration gradient. Thus, trace elements in free hydrophobic form (Cu^{2+}, Zn^{2+}, Cd^{2+} ...) easily cross membranes by membrane proteins or lipids binding.

2. Facilitated diffusion is another form of passive transport. It occurs through membrane structures such as pores, canals, aquaporins or permeases... (Buffle and DeVitre, 1993). Facilitated diffusion concerns hydrophilic and complex forms of chemical elements.

3. Active transport, a consuming energy process (adenosine triphosphatase -ATPase- or proton pump) where chemicals can pass through plasma membrane against concentration gradient. Thus, the apoplastic molecules bind to a transmembrane protein known as ion pump or transporter which releases it in the cytosol thanks to energy consumption. For example, copper has specific membrane transporters (Pfeil et al., 2014) like the family of Heavy Metal ATPase (HMA; Aguirre and Pilon, 2016).

Arsenic, as a non-essential element for plants, has neither specific and selective transporters nor pathways (Li et al., 2016). Arsenate for example, uses phosphate transporters (Farooq et al., 2016). Indeed, in solution, arsenate oxyanion occurs in three forms corresponding to three different pKa (2.2, 7.0 and 11.5) similar to phosphate pKa (2.1, 7.2 and 12.7). This arsenate-phosphate promiscuity can lead to non-discrimination between these two ions at the phosphate transporter level (Meharg and Macnair, 1992; Mkandawire and Dudel, 2005; Rahman et al., 2007; Zangi and Filella, 2012) and so, arsenate transmembrane penetration in cells can be achieved via phosphate transport systems. Analogy between phosphates and arsenates is well documented (Elias et al., 2012; Tawfik and Viola, 2011). On the other hand, the entry of arsenite into roots is mediated by aquaglyceroporins like for example Lsi1 (OsNIP2;1) from the nodulin 26–like intrinsic membrane protein family (NIP) identified in the rice roots

(Ma et al., 2008). The different mechanisms adopted by plants to uptake arsenic are schematized in the Figure 5 (Clemens and Ma, 2016; Zhao et al., 2009).

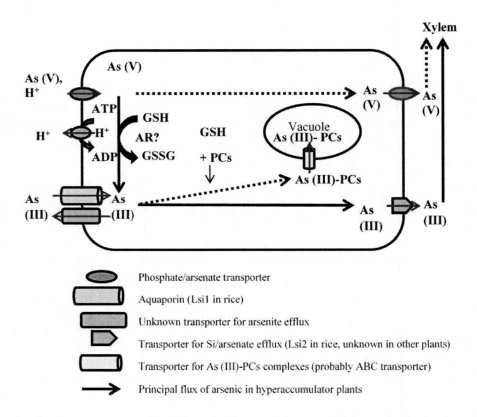

Figure 5. Diagram of putative and known mechanisms of arsenic uptake by plant cells (adapted from Zhao et al., 2009) (PCs: phytochelatins, As (III): arsenite, As (V): arsenate, GSH: reduced glutathione, GSSG: oxidized glutathione, AR: arsenic reductase).

3.4. Plant Arsenic Bioaccumulation

Bioaccumulation of metal and/or metalloids in plants is affected by several parameters such as plant species, chemical species and operating conditions (medium composition, duration and concentration of exposure).

In the literature, plant exposure to contaminants range from 1 to 32 days (Delmail et al., 2011; Rozentsvet et al., 2012; Srivastava et al., 2006; Swain et al., 2014; Xue et al., 2010) and usually, in laboratory conditions, occurs in synthetic media with a defined composition of nutrients (micro- and macroelements) (Delmail et al., 2011; Krayem et al., 2016a; Monnet et al., 2006; Ngayila et al., 2007) or polluted natural samples (Mishra et al., 2006). The difficulty to control the composition of the latter constitutes the main drawback. In order to study the plants ability to absorb and translocate xenobiotics from root to shoot, bioconcentration factor (BCF) and translocation factor (TF) are used. BCF is a key index to evaluate the plant's ability to accumulate pollutants in its tissues whereas TF evaluates the range of pollutant transfer from underground to above-ground parts of a plant (Liu et al., 2014).

$$TF=[M]shoots/[M]roots$$

[M]shoots is the concentration of metal or metalloid in shoots in $\mu g/g$ FW (Fresh Weight)

[M]roots is the concentration of metal or metalloid in roots in $\mu g/g$ FW.

$$BCF=[M]plants/[M]water$$

[M]plants is the concentration of metal or metalloid in plants (shoots or roots) in $\mu g/kg$ FW

[M]water is the concentration of metal or metalloid in water in $\mu g/L$.

The terrestrial fern *Pteris vittata* was the first arsenic hyperaccumulator identified in plant kingdom presenting a bioaccumulation factor of 160 when grown at 6 mg As.kg^{-1} of soil (Ma et al., 2001; Robinson et al., 2006). However, in the literature, several sources recognize macrophytes as potential strong metal bioaccumulators but this absorption capacity depends on macrophyte species, xenobiotic concentration and speciation, exposure duration and target organ. In the case of As, many aquatic macrophytes are demonstrated to perform the accumulation of this toxic element (Robinson et al., 2006; Shaibur et al., 2006; Van Den Broeck et al., 1997; Xue and Yan, 2011; Yabanli et al., 2014; Zhang et al., 2011). For example, an *Hydrilla verticillata* exposure to 0.002 µM As (V) during 4 days gives a BCF of 4 000 in root organs (Xue and Yan, 2011). After exposure to 10.6 mg As/kg sediment, BCF reaches 1.5 in roots and 0.045 in stems of *Najas marina* (Mazej and Germ, 2009). In *Myriophyllum alterniflorum* exposed to 1.6 µgAs (V).g^{-1} FW during 21 days, our team observed an increased BCF of 17% and 11% in roots compared to shoots in oligotrophic and eutrophic conditions respectively (Krayem et al., 2016a). This reduction in As translocation from roots to shoot could be due to phytochelatin assisted sequestration in roots vacuoles (Table 2). Also, an important parameter influencing significantly the As absorption by plants is the chemical composition of water in the aquatic environment. For example, As accumulation decreases with an increase of phosphate concentration as observed in *Lemna gibba* (Mkandawire et al., 2004). Actually, for a medium containing 133 µM arsenate and 0.37 µM phosphate, 1350 mg/kg FW arsenic is accumulated, whereas with 412 µM phosphate, arsenic accumulation reached only 971 mg/kg FW. In the same way, Rahman et al. (2008) noted a decrease of 68% of As concentration in *Spirodela polyrhiza* with the increasing concentrations of phosphate from 0.02 µM to 500 µM in the culture medium supplemented with 4 µM of arsenate. Previous results obtained in our laboratory (Krayem et al., 2016a) evidenced a 30% reduction of arsenic concentration in *Myriophyllum alterniflorum* exposed to 1.6 µgAs(V).g^{-1} FW during 21 days with the increase of phosphate concentration from 1 µM to 10 µM in the medium. Moreover, Tripathi et al. (2014) showed a double accumulation of arsenite (1 190 mg/g DW) and arsenate (645 µg/g DW) in *Najas indica* exposed to 246 µM arsenate combined with 18.6 µM arsenite. Despite the 5-fold higher dose of arsenate than arsenite, higher arsenite accumulation is due to phosphate competition (0.01 M). What is more, it is well demonstrated that arsenic preferentially accumulated in roots than in shoots (Mazej and Germ, 2009; Vromman et al., 2011; Yabanli et al., 2014). Our team did a similar observation on *Myriophyllum alterniflorum* where As concentration reached 110 µg/g and 160 µg/g dry weight (DW) in roots and 90 µg/g and 100 µg/g DW in shoots, in eutrophic and oligotrophic conditions respectively after an exposure to 1.3 µM of As (V) during 21 days (Krayem et al., 2016b). This differential accumulation of arsenic in roots (where As is directly detoxified by phytochelatins before sequestration in vacuoles) and shoots constitutes a protective mechanism of active physiological and developmental processes (like photosynthesis and cell division at the apical meristem).

Finally, as a recap of As (V) absorption, translocation and accumulation mechanisms previously described, a diagram is presented in Figure 6.

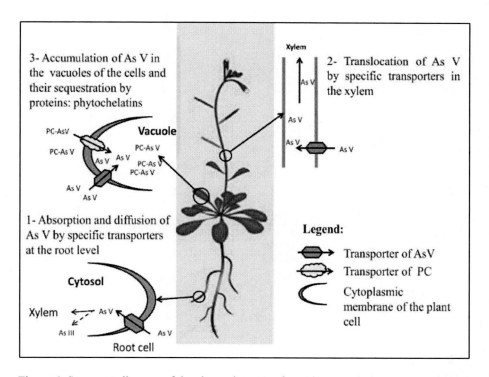

Figure 6. Summary diagram of the absorption, transfer and accumulation system of As in plants (modified from Clemens, 2001; PC: phytochelatins).

Table 2. Arsenic distribution in some aquatic plants

Studies	Plant model used and applied dose of arsenic	[As] in total plants in µg/g of DW	[As] in stems in µg/g of DW	[As] in roots in µg/g of DW
Mazej and Germ, 2009	*Najas marina* (4.87-20.3 mg/kg of sediment, <1 µg/L in water, field study)		<2	15.9
	Potamogeton lucens (4.87-20.3 mg/Kg of sediment, <1 µg/L in water, field study)		<2	12.7
Ozturk et al., 2010	*Nasturtium officinale* R. Br.), 50 µM of As (III), Hoagland's medium	1012		
Duman et al., 2010	*Lemna minor* 64 µM of As (III) and As (V) separately, 10% Hoagland's solution	17 408 and 8674 respectively		
Xue and Yan, 2011	*Hydrilla verticillata*, 20 µM As (III), synthetic fresh water	821		
Xue and Yan, 2011	*Hydrilla verticillata*, 20 µM As (V), synthetic fresh water	715		
Zhang et al., 2011	*S. polyrhiza*, 320 µM As (V), hydroponic culture solution	999±95		
Srivastava et al., 2013	*Hydrilla verticillata*, 500 µM of As (V), Hoagland's solution (pH 6.5)	568		
Yabanli et al., 2014	*Myriophyllum spicatum* (0.01-0.013 µM of As, field study)		0.32	15.30
Tripathi et al., 2014	*Najas indica* (250 µM of As (V), in 10% Hoagland's solution	645		
Tripathi et al., 2014	*Najas indica* (50 µM of As (III), in 10% Hoagland's solution	1 190		
Olszewska et al., 2016	*P. amphibia, P. pectinatus, E. nuttallii, M. spicatum, and Chara spp* in 2 sites: Site 1 (87 mg kg−1) and 6 sites across the lake of Kinghorn Loch scotland (mean as concentration: 160 mg kg−1)	A mean of 8.8 in the five taxa		
Krayem et al., 2016a	*Myriophyllum alterniflorum* 1.3 µM of As (V), synthetic vienne medium eutrophic and oligotrophic		90 and 100 in eutrophic and oligotrophic conditions	110 and 160 in eutrophic and oligotrophic conditions

4. PHYTOTOXICITY OF ARSENIC

Once in the cytoplasm of plant cells, chemical elements interfere with plant metabolism processes. Non-essential elements like arsenic are directly toxic and can drastically alter plant metabolism. Toxicity symptoms can be directly visible by affecting plant morphology (chlorosis, necrosis...) and/or can be detected by detailed analyses at histological, physiological, biochemical or genetic level.

4.1. Alteration of Growth and Development

Plant growth corresponds to mitotic divisions occurring in meristems, followed by cell elongation (Arduini et al., 1995). Beyond this cellular aspect of plant development, three phases can be distinguished in plant growth: a phase of latency, a proliferative phase and a phase where proliferation slows down ending by senescence. The third phase occurs earlier during periods of stress and the previous growth reduction observed constitutes the first stress response as seen in tomatoes, wheat, maize and beans (Adriano, 2001; Baccouch et al., 1998; Chaoui et al., 1997; Lin et al., 2007; Sandalio et al., 2001; Sobkowiak and Deckert, 2003). Among the biological effects of metal pollution, the inhibition of plant growth and development has been demonstrated in many organisms (Mench and Bes, 2009). For example, in the case of wheat seedlings exposed from 0.066 to 0.266 mM As during 7 days, Li et al. (2007) showed a decrease in germination and in stem and root growth. On this last point, it is important to note that roots are the first structures damaged given their position as the entry point of chemical elements (Muller et al., 2001). Growth reduction is a common response to As exposure as already observed in other crops like bean (Stoeva et al., 2005), lettuce (Gusman et al., 2013) and rice (Rahman et al., 2007). In addition, growth inhibition, chlorosis, leaf necrosis and delayed flowering are other consequences of the As toxicity in *Zea mays*, *Brassica napus*, *Helianthus annuus* and *Lolium perenne* exposed to 37.33 µM As during 4 months (Gulz et al., 2005).

Concerning aquatic macrophytes, *Myriophyllum alterniflorum* exposed to 1.6 µgAs (V).g^{-1} FW during 30 days in eutrophic condition, presented a 43% reduction in stem length and a 91% reduction in ramification length (Krayem et al., 2016b). A 15% reduction in stem length and a 40% reduction in root length were also noted after an exposure to 143 µgAs.g^{-1} FW during 21 days in an aquarium system (Krayem, 2015).

4.2. Physiological Alterations

4.2.1. Alteration of Photosynthesis and Photosynthetic Pigments

Photosynthesis is a vital mechanism in terrestrial and aquatic plants. During photosynthesis, two phases are distinguished: (i) the photosynthetic electron transport chain from water to nicotinamide adenine dinucleotide phosphate (NADP) and (ii) the Calvin cycle also known as Calvin-Benson-Bassham cycle (Keller, 2015; Figure 7).

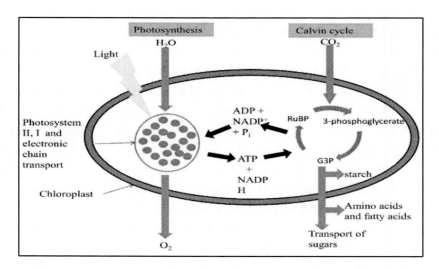

Figure 7. The different phases of photosynthesis and the Calvin cycle in plant cell (adapted from Keller, 2015) (NADP: nicotinamide adenine dinucleotide phosphate, NADPH: dihydronicotinamide adenine dinucleotide phosphate; ADP: adenosine diphosphate; RuBP: ribulose 1,5-biphosphate; ATP: adenosine triphosphate; G3P: 3-phosphoglycerate).

From a biochemical point of view, many factors can affect photosynthesis like light intensity, trophic level or temperature (Nielsen, 1993) but metallic stress can also have some important impacts on the enzymes of photosynthesis and respiration and on the photosynthetic pigments (Poschenrieder and Barceló, 2004; Sandalio et al., 2001; Sanità di Toppi and Gabbrielli, 1999). Excessive photon energy induces overproduction of excited electrons and thus over-excitation of the electron transport chain. This electronic overload leads to reactive oxygen species (ROS) formation inducing an oxidative stress (Sharma et al., 2012). In the same way, xenobiotics can affect photosynthesis. Metal can induce lipoperoxidation of chloroplasts membrane system (Pinto et al., 2003). Moreover, metal can also destroy the electron transport chain resulting in ROS production and leading to a reduction in photosynthetic activity (Chugh and Sawhney, 1999). In laboratory experiments, Díaz et al. (2013) found that the net photosynthesis rate decreased in the aquatic moss *Fontinalis antipyretica* exposed to an As concentration range from 0.00133 μM to 133.33 μM during 22 days. Moreover, Ritchie and Mekjinda (2016) showed that arsenic at 0.1 mol.m^{-3} for 24 h was able to decrease more than 90% the maximum gross photosynthesis of the small aquatic angiosperm *Wolffia arrhiza*. As essential actors of the photosynthesis, the photosynthetic pigments (Chl a, b and carotenoids) absorb light and turn light energy into chemical energy in thylakoid membranes of chloroplasts. However, as previously said, trace metals and metalloids induce ROS production, resulting in pigment damages (MacFarlane, 2003). Indeed, Mg^{2+} from the tetrapyrole (=porphyrin) ring of chlorophyll molecules can be substituted by metals like Cu^{2+}, Cd^{2+} and Zn^{2+} (Prasad, 1998). Among photosynthetic pigments, chlorophyll b is usually more sensitive than chlorophyll a, while carotenoids are more resistant to metal/metalloid stress. Actually, carotenoids protect the photosynthetic system from excessive light and from trace metals responsible of ROS production as they act as ROS scavengers (Gill and Tuteja, 2010). Specifically on As studies, several authors reported the toxic effect of this compound on photosynthesis and chlorophyll pigments in different plant species (Rahman et al., 2007; Stoeva et al., 2005; Stoeva and Bineva, 2003). In terrestrial species, Mascher et al. (2002) evidenced a decrease in chlorophylls a, b and carotenoids contents in *Trifolium pratense* after exposure to 10 to 50 mgAs.kg^{-1} soil during 10 weeks. In *Pteris ensiformis,* a 10 day exposure to 133 and 264 μM arsenic-supplemented medium significantly reduced chlorophyll a and b contents (Singh et al., 2006).

For aquatic macrophytes species, a decrease of chlorophyll a and b contents was demonstrated in the case of 6 day exposure of *Lemna minor* to Hoagland medium supplemented with arsenate from 1 μM to 68 μM (Duman et al., 2010). Ozturk et al. (2010) obtained similar results with *Nasturtium officinale* exposed for 7 days to arsenite from 1 μM to 52 μM in Hoagland medium. Srivastava et al. (2013) also noted a 24% decrease in pigment content and net photosynthesis activity in *Hydrilla verticillata* exposed 96 h to 43 μM of arsenate. Moreover, in watermilfoil exposed during 14 days to 1.6 μgAs (V).g^{-1} FW in eutrophic conditions, a 34% decrease of net photosynthesis and 39% decrease of respiration were observed (Krayem et al., 2016b). This decrease was most certainly due to a 22%, 11% and 17% reduction in chlorophyll a, b and carotenoids, respectively. In contrast, some results obtained by Rofkar et al. (2014), showed in the aquatic plant *Azolla caroliniana* an increase in anthocyanin content and a 30% decrease in growth after 8.1 μM As (V) exposure during 14 days. These pigments are antioxidant compounds, sometimes produced in response to metal stress. Moreover, in *Vallisneria gigantea* and *Azolla filliculoides* exposed to 0.0266 mM of arsenate, Iriel et al. (2015) observed a fluorescence increase in the range of 400-500 nm related to increase of carotenoid and flavonoid contents. These compounds are an integral part of arsenic defense system. During this study, the authors also evidenced an As-induced decrease in the photosystem II (PSII) quantum yield efficiency which led to a reduced photosynthesis. In parallel, they observed a carotenoid content decrease after 5 and 10 days of 264 μM As exposure. For As concentration of 0, 1.44, 7.2 and 36 mg As (V).kg^{-1} dry sediment, Gross et al. (2016) obtained a chlorophyll and carotenoid contents decrease in *Myriophyllum spicatum* (14 days exposure). Finally, from the abundant examples of As impact on photosynthetic pigments, a study on young leaves of *Ceratophyllum demersum* exposed for 4 weeks to 25 μM As under low phosphorous (P) levels (1 µM), presented 86% reduction of chlorophyll content compared to control (Mishra et al., 2014). In fact, the formation of an ADP-As (V) complex occurred as arsenic and particularly arsenate could replace phosphates. This irreversible complex inhibits ATP production originating from both respiration in the mitochondrial electron transport chain and photosynthesis in thylakoids (Finnegan and Chen, 2012). Moreover, chlorophyll synthesis occurred from coproporphyrinogen III (Copro III), a precursor shared by both heme and chlorophyll biosynthetic pathways. A two steps transformation leads to protoporphyrin IX from Copro III. Thus, insertion of Fe or Mg into protoporphyrin IX results in heme or Mg-protoporphyrin synthesis, respectively. In another study with *C. demersum* treated for 2 weeks with 0.5 µM As, a strong decrease of all the chlorophyll precursors contents was also observed (Mishra et al., 2016). These authors suggested that As could inhibit and block the tetrapyrole pathway leading to weak chlorophyll synthesis efficiency.

4.2.2. Alteration of Osmotic Potential (Ψs)

The osmotic potential (Ψs) depends on the solute concentration of the media where the plant is immersed. Osmotic potential is defined through relationship between plant water potential (Ψ) and pressure potential (Ψp) as follows:

$$\Psi = \Psi p + \Psi s$$

Pressure potential represents excess of hydrostatic pressure compared to atmospheric pressure. The osmotic potential (Ψs) is also known as solute potential. An increase in vacuolar or cytoplasmic solute concentration leads to a decrease in osmotic potential (which becomes more negative; Hopkins, 2003). Not only metal stress but also increase in nitrate, phosphate or ammonium concentrations can lead to a decrease in osmotic potential. Indeed, this event is due to the positive correlation between ammonium concentration and plant cytosolic accumulation of amino acid (Cao et al., 2007) and inorganic anion (Kronzucker and Britto, 2002).

Toxic elements such as As are also known to induce a decrease in osmotic potential. It is particularly the case for terrestrial species *Atriplex atacamensis* (xerohalophyte), which saw a decrease in osmotic potential after an exposure to 100 µM As during 28 days (Vromman et al., 2011). Similarly, Stoeva et al. (2005) evidenced that exposure to 5 mgAs.kg^{-1} soil during 5 days induced a decrease in root osmotic potential in two *Phaseolus vulgaris* varieties.

For aquatic macrophytes, a 54% and 42% increase in amino acids and proteins contents, respectively, was noted by Tripathi et al. (2014) after 2 days-exposure of *Najas indica* to increasing arsenate concentrations (from 0 to 0.244 mM). This increase in amino acid and protein concentrations could induce a decrease in osmotic potential. Moreover, Srivastava et al. (2013) showed that 5 mM As (V) exposure during 96h reduced water use efficiency and increased transpiration in *Hydrilla verticillata*. This reduction in water use efficiency can increase solutes concentration in plant and therefore decrease the osmotic potential. Finally, a 7% and 4% decrease in osmotic potential of *Myriophyllum alterniflorum* exposed to 1.6 µgAs (V).g^{-1} FW was also observed in oligotrophic and eutrophic conditions respectively (Krayem et al., 2016b).

4.3. Oxidative Stress and Membrane Peroxidation

Oxidative stress is related to the production of a number of free radicals, commonly referred to Reactive Oxygen Species or ROS. ROS cause intracellular degradation in plants by altering the cell metabolism and the gene expression. The photosynthetic electron transport chain is the main source of ROS in the organism tissues (Gill and Tuteja, 2010). There are two categories of ROS: primary and secondary. Primary ROS is formed by superoxide anion (O_2*), hydrogen peroxide (H_2O_2), hydroxyl radical ($*OH$), hydroperoxyl radical ($*OOH$) and singlet oxygen ($^1O_2$). The generation of the superoxide anion (O_2*^-) takes place in mitochondria, peroxisomes, glyoxysomes or by the Mehler reaction (1) in chloroplasts of plant cells:

$$O_2 + e^- \rightarrow O_2^{*-} \text{ (Mehler reaction)} \qquad (1)$$

Hydrogen peroxide (H_2O_2) is produced by the dismutation of the superoxide anion (O_2^{*-}) in the presence of superoxide dismutase (SOD). The hydroxyl (*OH) radical is produced by the Fenton reaction in the presence of the Cu^{2+} and Fe^{2+} ions according to the Fenton reaction (2) or according to the Haber-Weiss (3) reaction:

$$O_2^{*-} + Fe^{3+} \rightarrow O_2 + Fe^{2+}$$
$$H_2O_2 + Fe^{2+} \rightarrow Fe^{3+} + HO^- + \text{*OH (Fenton reaction)} \qquad (2)$$

$$O_2^{*-} + H_2O_2 \rightarrow \text{*OH} + HO^- + O_2 \text{ (Haber-Weiss reaction)} \qquad (3)$$

These oxygen radicals, apart superoxide anion (O_2^{*-}), can spread very easily through cell membranes and cause oxidative damage to the carbohydrates, lipids, amino acids, proteins and nucleic acids (da Silva et al., 2013; Figure 8).

Sometimes, these oxidative damages due to primary ROS (especially the hydroxyl radical or the superoxide anion) contribute to the formation of secondary ROS. Among these radicals, there are alkyl R*, alkoxyl RO* and peroxyl ROO *. For example, the latter is formed by the action of a primary ROS on the chain of an unsaturated fatty acid, which in turn transforms it into a ROO*. This mechanism corresponds to lipid peroxidation or lipoperoxidation (Birben et al., 2012). The degree of this peroxidation is generally estimated by measuring the concentration of malondialdehyde (MDA). This organic compound (chemical formula $CH_2(CHO)_2$) resulting from the degradation of polyunsaturated fatty acids by ROS constitute an excellent biomarker of oxidative stress (Xing et al., 2010). Indeed, metal-metalloid stress by lead (Mishra et al., 2006), iron, copper or arsenic is known to increase the plant cell concentration of this aldehyde.

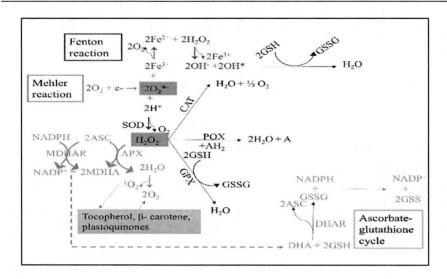

Figure 8. Different forms of reactive oxygen species and their production in the plant cell (adapted from da Silva et al., 2013) (CAT: catalase; POX: peroxidase, APX: ascorbate peroxidase; MDHA; mono-dehydro-ascorbate; MDHAR; mono-dehydro-ascorbate reductase NADP: nicotinamide adenine dinucleotide phosphate NADPH: dihydronicotinamide adenine dinucleotide phosphate; GPX cytochrome oxidase; GSH: reduced glutathione; GSSG: oxidized glutathione).

Arsenic is also demonstrated in several studies (Ahsan et al., 2008; Hasanuzzaman et al., 2015; Mallick et al., 2011) as an inducer of the production of ROS: hydrogen peroxide (H_2O_2; Barchowsky et al., 1996; Mascher et al., 2002; Pisani et al., 2011; Singh et al., 2006; Stoeva et al., 2005), hydroxyl radical (HO*) and superoxide anions ($O2^{*-}$; Srivastava et al., 2013; Yamanaka et al., 1997). After the activation of these species, As stress is also well known to activate the MAPK (Mitogen-activated protein (MAP) kinases) cascade. MAPK are serine/threonine/tyrosine specific signaling proteins which are responsible to signal conversion from the receptors into cellular responses (Islam et al., 2015). Another damage caused by As is the peroxidation of membrane lipids (Shri et al., 2009) which contributes to the generation of lipid peroxides or MDA (Chen et al., 2015; Duman et al., 2010; Singh et al., 2006). For aquatic macrophytes, Krayem et al. (2016b) found an increase of 62% of H_2O_2 and 35% of MDA in the tissues of *Myriophyllum alterniflorum* exposed for 21 days to 1.6

μgAs (V).g^{-1} FW (*in vitro* experiment with eutrophic conditions). Similarly, Chen et al. (2015) demonstrated that arsenic at 6.66 μM and 26.6 μM was able to increase of about 80% the MDA content in *Vallisneria natans* after 96 h of treatment and Duman et al. (2010) showed an increase of about 49% in the MDA content after 64 μM arsenate and 6 days of treatment in *Lemna minor* plants. Tripathi et al. (2014) showed that the MDA content of the As (V) and As (III) treated plants increased compared to their respective controls with a maximum induction of 80% and 100% respectively after an exposure for 7 days of aquatic macrophyte *Najas indica* to 250 μM of As (V), and 50 μM of AsIII in two separate experiments. In another study performed by Ozturk et al. (2010), MDA content increased to a highest value of 86% in comparison with the control after application of 10 μM As(III) in the tissues of *N. officinale* over 7 days. Thus, the reactive oxygen species would be partly responsible for the toxicity of arsenic.

4.4. Genotoxicity

Plant meristem and particularly root meristems are very sensitive to xenobiotics-induced genotoxic effects. Therefore, root meristems are commonly used for toxicity evaluation through cytological studies. The micronucleus test or *in vitro* MCN test has been used since 1938 for the detection of external toxicity on plant root meristems (Zaka et al., 2002). Micronuclei are chromosome fragments or entire chromosome detached from mitotic spindle during cell division and transformed in small nucleus-like structure (Figure 9; Souguir et al., 2008; Zaka et al., 2002).

It is well known that xenobiotic induces genotoxic effects in roots leading to a decrease in mitotic index, formation of micronuclei, induction of chromosomal aberrations and DNA and RNA alterations (Béraud et al., 2007; Borboa and Torre, 1996; Fusconi et al., 2006; Ünyayar et al., 2006). For example in terrestrial species, Gichner et al. (2004) and Liu et al. (2003) studied the effects of cadmium on the root meristems of *Nicotiana tabacum* and *Allium sativum* respectively. These authors noted an increased DNA damage leading to micronuclei formation and a low mitotic index. In addition these aberrations could not inhibit the root elongation (Jiang et al., 2000; Seregin and Ivanov, 2001). Copper for example, can also be genotoxic and cause DNA damage resulting in chromosomal fragmentation leading to micronuclei formation (Souguir et al., 2008; Yıldız et al., 2009). Fusconi et al. (2006) observed an increased frequency of micronuclei formation in roots of *Vicia faba* and *Pisum sativum* exposed to 2.5 mM CuSO$_4$ during 42h. These genotoxic effects are probably related to ROS formation released during metal stress (De Marco et al., 2002; Yi and Si, 2007). In addition, as shown in several *in vitro* and *in vivo* studies, arsenic also induces chromosomal alterations, DNA damage and high recombination rates, exchange between chromatids and micronucleus formation in humans, animals and plants (Anjum et al., 2012; Gebel, 2001; Gonsebatt et al., 1997). Arsenic genotoxic effects were studied in two terrestrial plant roots: *Vicia faba* and *Allium cepa* (Wu et al., 2010; Yi et al., 2007). In these studies, arsenic induces micronuclei formation in the roots with a frequency increasing with arsenic concentration. *Vicia faba* root cells exposed from 4 to 133 μM As appeared more sensitive than those of *Allium cepa* exposed from 13.33 to 400 μM As.

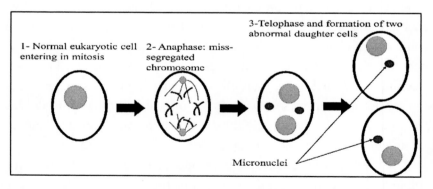

Figure 9. Formation of micronuclei in eukaryotic cells (adapted from Meneguetti and da Silva, 2012).

For aquatic macrophytes, few studies related to micronuclei formation are available. In roots *Elodea canadensis* exposed from 8 to 17 mgCu/kg sediments in a field experiment associated with other chemical pollutants

(Zn, Cr, Pb, Ni and Cd), an increased percentage of abnormal cells (from 1.8% to 16.4%) was evidenced by Zotina et al. (2015). In young mature leaves of *Ceratophyllum demersum* exposed to 1 µM As during 2 weeks, arsenic was predominantly accumulated in the nucleus of epidermal cells (Mishra et al., 2016). As said previously, arsenic could replace phosphorous in nucleic acid synthesis leading to deleterious effects.

5. DEFENSE AND TOLERANCE

In response to the toxic elements and to reduce toxicity caused by their adsorption then their absorption by plants, two systems are activated: detoxification system and antioxidant defense system.

5.1. Detoxification System

Detoxification mechanisms can be of extra and/or intracellular origin and vary according to plant species (Hall, 2002). On one hand, extracellular strategies include all mechanisms allowing contaminants collection and excretion thus avoiding *in planta* accumulation. As an example, this strategy occurs through CPx type ATPase, an ATPase ion pump allowing copper transport outside the cells (Hall, 2002). On the other hand, intracellular strategies include complexation of contaminants by macromolecules and vacuolar sequestration *via* carriers and chaperone proteins (Clemens, 2001). It is also possible *via* detoxification through chelation with ligands like amino acids (histidine, cysteine, glycine, etc.), organic acids (malic acid, citric acid, etc.), vitamin E or α-tocopherol, carotenoids, flavonoids, phenols, phytochelatins (PC; known as specific biomarkers of metal pollution; Srivastava et al., 2013; Yin et al., 2002) and polyamines. Recently, a class of amines known to play an important role in plant growth and development has been studied in response to various environmental stresses: polyamines, and particularly putrescine (Bouchereau et al., 1999; Yang et al., 2010).

5.1.1. Phytochelatins

Plant cells have developed a protection mechanism against metal stress: metal ions entering the cytosol are immediately complexed and inactivated. This mechanism involves peptide structures called phytochelatins (PCs) which are small heavy metal binding polypeptides rich in cysteine. PCs are structurally related to glutathione and represented as chain of oligomers (γ-Glu-Cys)$_n$-Gly, (n = 2-11). PCs are present not only in plants, but also in fungi and other organisms (Cobbett, 2000; Yadav, 2010; Zenk, 1996) (Figure 10).

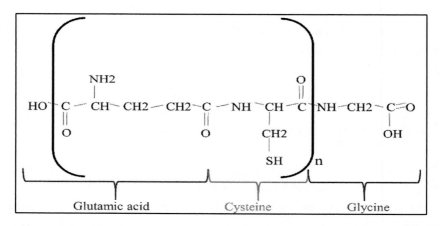

Figure 10. Structure of a phytochelatin (2<n<11) (adapted from Steffens, 1990).

Phytochelatine synthase (PCS), a γ-glutamylcysteine dipeptidyl transpeptidase (EC 2.3.2.15), catalyzes the transpeptidation of γ-Glu-Cys unit from one GSH molecule to another to form PC$_2$. After the accumulation of sufficient levels of PC as substrate, PCS catalyzes the formation of PC$_{n+1}$ by the transpeptidation of a γ-Glu-Cys unit from GSH to a PC (PC$_n$) molecule (Ha et al., 1999; Rea et al., 2004). Some Fabales and Poaceae synthetize PC containing β-alanine, serine or glutamic acid in C-terminus. The biological role of these variants is still unclear and must be elucidated (Inouhe, 2005; Zenk, 1996). Increased PC synthesis occurs during xenobiotic exposure (Zenk, 1996). PCs offer chelating properties leading to metal binding with thiol groups. The metal-PC complex could be sequestered in vacuoles thus reducing the metal toxicity (Satofuka et al., 2001). High concentrations of essential elements like copper and non-essential contaminants such as arsenic, lead and cadmium stimulate PC synthesis and thereafter are submitted to vacuolar sequestration as PC-ETM complexes (Grill et al., 1989). Thus, metal and metalloids pollution increase clearly cytosolic PC concentration, also in the case of As contamination (Chandrakar et al., 2016). Indeed, once in roots, arsenite or arsenate (after reduction to arsenite by arsenate reductase (EC:1.20.4.1)), are complexed by thiol groups of PC and then sequestered in root cell vacuoles (Dhankher et al., 2006; Farooq et al., 2016). PC synthesis was firstly reported in terrestrial species *Rubia tinctorum* exposed to various metals: Hg^{2+} (10 µM), Ga^{3+}, Ln^{3+}, Pb^{2+}, Zn^{2+}, Ag^+ (100 µM), As^{3+}, As^{5+}, Cd^{2+}, Cu^{2+}, Ni^{2+}, Pd^{2+} and Se^{4+} (1000 µM) in Murashige and Skoog medium (Maitani et al., 1996). In *Cicer arietinum* exposed to 20 µM As during 2 days, PC (PC_2, PC_3) synthesis occured in the roots (Gupta et al., 2004). In the aquatic macrophyte *Wolffia globosa,* 74% of accumulated As was complexed by phytochelatins (PC_2 and PC_3) and sequestered in vacuoles (Zhang et al., 2012). In *Myriophyllum alterniflorum* exposed 21 days to 1.6 µgAs.g^{-1} FW in eutrophic and oligotrophic conditions, our team observed PC_2 and PC_3 accumulation at 3 days with a maximum at 7 days followed by a decrease until 21 days (Krayem, 2015). During this experiment, PC_2 reached higher concentration than PC_3. Similar results were obtained previously in *Najas indica* exposed from 10 to 250 µM arsenate (Tripathi et al., 2014). A 1.5-fold increase in PC_2 (from 157 to 245 nmol.g^{-1} FW) and a 2-fold increase in PC_3 contents (from 42.3 to 92.64 nmol.g^{-1} FW) were noted with the increase in As (V) concentration. In all treatments, PC_2 surpass PC_3 contents. PCs contents are also positively correlated to As III content and exposure. Once again, compared to control, the increase in PC_2 (34-fold, 209 nmol.g^{-1} FW) surpasses the increase in PC_3 (12-fold, 50 nmol.g^{-1} FW; Tripathi et al., 2014).

5.1.2. Polyamines

In most organisms, polyamines (PAs) are omnipresent biogenic amines implicated in various cell functions (Hussain et al., 2011). PAs are very low-molecular-weight aliphatic nitrogen compounds and positively charged at physiological pH (Bouchereau et al., 1999; Takahashi and Kakehi, 2010). Three major free or single PAs are distinguished: putrescine (Put), spermidine (Spd) and spermine (Spm) (Kumar et al., 1997; Tiburcio et al., 1997). Conjugated polyamines are PAs bound to various molecules (Martin-Tanguy, 2001; Tiburcio et al., 1997). In plants, PAs biosynthesis is well documented in the literature. Putrescine (Put) synthesis occurs either directly through ornithine decarboxylase (ODC, EC 4.1.1. 17) pathway from ornithine or indirectly through arginine decarboxylase (ADC, *EC* 4.1.1.19) pathway from arginine with agmatine and carbamoylputrescine intermediates. Agmatine decarboxylation involves agmatine iminohydrolase (AIH also known as agmatine deiminase, EC 3.5.3.12) and N-carbamoylputrescine amidohydrolase (CPA also known as N-carbamoylputrescine amidase, EC 3.5.1.53). Spermidine (Spd) and spermine (Spm) synthesis occurs through successive addition of aminopropyl on Put and Spd respectively. These reactions are catalyzed by Spd synthase (SPDS, EC 2.5.1.16) and Spm synthase (SPMS, EC 2.5.1.22) (Figure 11) (Hussain et al., 2011; Kusano et al., 2007; Martin-Tanguy, 2001).

PAs are involved in numerous cellular processes such as cell division, cell elongation, root, flower and fruit growth, replication, transcription, translation, ribosome biogenesis, somatic embryogenesis, preservation of photosynthesis and programmed cell death (Bertoldi et al., 2004; Bouchereau et al., 1999; Faivre-Rampant et al., 2000; Groppa and Benavides, 2007; Navakoudis et al., 2003). PAs are also involved in plant responses to a wide range of environmental stresses including high temperatures (Bouchereau et al., 1999; Groppa and Benavides, 2007; Hussain et al., 2011; Königshofer and Lechner, 2002; Yang et al., 2010) or NH_4^+ pollution (Shen et al., 1994). Moreover, PAs concentrations vary during copper (Groppa et al., 2003; Pirintsos et al., 2004; Sharma, 2006; Wettlaufer et al., 1991) and arsenic stress (Mascher et al., 2002; Vromman et al., 2011). PAs accumulation is assumed to limit stress as PAs could trap oxygen free radicals or inhibit their production (Balestrasse et al., 2005; Groppa et al., 2003; Takahashi and Kakehi, 2010).

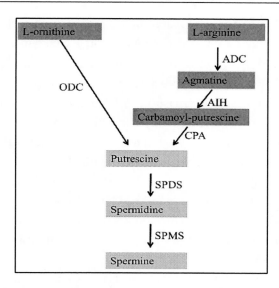

Figure 11. Synthetic pathways of polyamines in plants (modified from Takahashi and Kakehi, 2010). Abbreviations: ADC, arginine decarboxylase; AIH, agmatine iminohydrolase; CPA, N-carbamoylputrescine amidohydrolase; ODC, ornithine decarboxylase; SPDS, spermidine synthase; SPMS, spermine synthase.

5.2. Anti-Oxidative Defense System

Three types of enzyme are considered as key enzymes of the antioxidant defense system to eliminate ROS: superoxide dismutase (SOD, EC 1.15.1.1), catalase (CAT, EC 1.11.1.6) and various peroxidases (Px, secreted class III plant peroxidases (EC 1.11.1.7); superfamily including L-ascorbate peroxidase, APx, EC 1.11.1.11 and glutathione peroxidase, GPx, EC 1.11.1.9) (Cosio and Dunand, 2009; Valério et al., 2004)). SOD is a family of metalloenzyme (Cu, Zn, Mn and Fe) involved in dismutation of superoxide anion to hydrogen peroxide and dioxygen. CAT dismutates H_2O_2 in water and oxygen. APx or GPx could also scavenge hydrogen peroxide (Figure 8). The antioxidant defense induced by metals (Cu and Cd are the most studied) and metalloids (especially As) is quite important in terrestrial and aquatic plants. For terrestrial plants, after As exposure, an increase in the levels of SOD, Px, APx and glutathione reductase (GR) in rice grains and maize plants has been reported (Requejo and Tena, 2005;

Shri et al., 2009). For aquatic macrophytes, Schröder et al. (2009) noted an increase in enzyme activity of GR, APX and Px enzymes in *Phragmites australis* exposed to 10 to 500 µM of lead, cadmium and arsenic on *Typha latifolia*. In this experiment, these authors also observed an increase in GR and Px activities with a decrease in APX activity in *Typha latifolia*. In *Najas indica*, Tripathi et al. (2014) showed different evolutions in APx, GPx, SOD and CAT activities during 7 days, according to arsenic speciation (III and V) and concentrations applied. In *Nasturtium officinale* exposed for 7 days to 5 µM As (III), SOD activity increased by 37% (Ozturk et al., 2010). In *Lemna minor,* SOD, CAT and APx, activities were higher with the increase in arsenic concentration (0, 1, 4, 16 and 64 µM), to reach a maximum threshold at 1-2 days and a minimum after 7 days (Duman et al., 2010).

A summary of defense system against metals and metalloids involving PCs and antioxidant is illustrated in Figure 12.

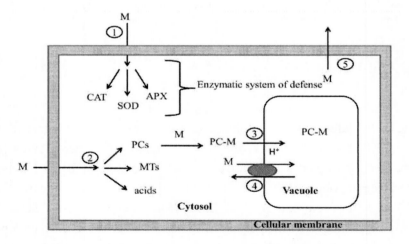

Figure 12. Plant-adapted enzymatic and non-enzymatic defense system to control pollutants in a plant cell (modified from Delmail et al., 2013; Hall, 2002) 1-Induction of the enzymatic defense system. 2-Detoxification by intracellular ligands. 3- Storage and sequestration of PC-M complexes in vacuole. 4-Accumulation of metals/metalloids in vacuole. 5-Exclusion of metals/metalloids outside cells. Abbreviations: M (trace element metal or metalloid); CAT (catalase); SOD (superoxide dismutase); APx (ascorbate peroxidase); PC (phytochelatin); MT (metallothionein).

6. WATER QUALITY BIOMONITORING

Water quality can be assessed through chemical and biological monitoring, generally by measuring the extent of environmental contamination and evaluating the impact of the pollutant on organisms, respectively. Chemical monitoring consists in water sampling, conservation and then laboratory analysis. However, continuous monitoring is limited by the sampling number and the representativeness of the sample is often uncertain. Moreover, for certain xenobiotics, the limit of quantification may be higher than the dose impacting the organisms (Lagadic et al., 1998). Recently, a new technique has been developed to determine the concentration of labile or bioavailable trace metal in the environment, even at low concentration: the Diffusive Gradient in Thin film (DGT). It consists of three main layers: a resin-impregnated with a polyacrylamide gel, a polyacrylamide gel diffusion layer and a filter membrane. Metal ions flow through diffusion layer and accumulate on the cation-exchange resin immobilized in the hydrogel (Conesa et al., 2010; Dočekalová et al., 2015; Zhang et al., 2001). DGT can provide information on the concentration of trace metals, metalloids like arsenic and nutrients such as nitrate, ammonium and phosphorus (Buzier et al., 2014; Dočekalová et al., 2015; Tandy et al., 2011; Zhang et al., 2001). DGT also allows an integrative sampling by implementing a device over a period of several days or even weeks. Thus, DGT is more representative of the pollutant concentration than a point sampling. However, this tool is still under development in order to facilitate its handling, the implementation conditions (duration of exposure, biofilm development, etc.) and data processing. DGT is unable to evaluate the effects of contaminants on living organisms. Therefore, finding methods to overcome the inadequacy of analytical methods is still a major concern.

6.1. Sentinel Species and Biomarkers

In environmental sciences, sentinel species can provide relevant information on the availability, types, quantities and effects of contaminants on the environment. They allow the detection of pollution when low concentrations are undetectable by analytical techniques or when integrative sampling is not applicable. A sentinel species is a sensitive species, a biological monitor with the ability to bioindicate one or more pollutants (Kaiser, 2001). Due to their sensitivity, these species are able to detect early environmental alteration before other species. Sentinel species represent a kind of alarm signal for the entire ecosystem (Hopkin, 1989; Wood, 2011). Sentinel species are characterized by: (i) easy handling, (ii) sufficient population coverage (without being invasive) and (iii) growing in known range of ecological conditions (Kaiser, 2001). Moreover, biology and ecology of the selected species must be well known and bioindicator ability confirmed by previous studies. A species may become a bioindicator able to evaluate the risks associated with xenobiotics if its biomarkers are affected. Many sentinel species are already used for water monitoring including aquatic mosses like *Sphagnum acutifolia, Fontinalis antipyretica, F. squamosa* (Besse et al., 2012), planktonic algae like *Microcystis, Scenedesmus, Chlorella, Hydrodictyon, Cladophora*, benthic marine algae like *Stigeoclonium and diatoms*, aquatic ferns like *Azolla, Salvinia*, and macrophytes (Ansari et al., 2016; Zhou et al., 2008).

To date, only few studies deal with macrophytes to detect pollution (*Lemna minor, Eichhornia crassipes, Spirodela polyrhiza, Typha latifolia*, etc.) and among them, immersed macrophytes are scarce (*Hydrilla verticillata, Ceratophyllum demersum, Myriophyllum quitense*). *Myriophyllum alterniflorum*, an european macrophyte from oligotrophic river could be a good indicator; it has already been studied in our laboratory (Chatenet et al., 2006; Delmail, 2011; Krayem, 2015; Monnet et al., 2005; Ngayila et al., 2007). Lagadic et al. (1998) present the well-known definition of a biomarker as "an observable and/or measurable change at the molecular, biochemical, cellular, physiological or behavioral level that reveals past or present exposure of an individual to at least one pollutant." Indeed, biomarkers represent biological responses to disturbances or contamination of the environment. Three types of biomarkers can be distinguished: biomarkers of exposure, biomarkers of effects and biomarkers of sensitivity/susceptibility (Lagadic et al., 1998;

Vasseur and Cossu-Leguille, 2003). Biomarkers of exposure like DNA damage, stress proteins or PC reflect the organism exposure to pollutant. Biomarkers of effect correspond to the effects of pollutants on the organism leading to modifications of biological functions like lipid peroxidation or increase in the antioxidant defense system. Biomarkers of sensitivity/susceptibility reflect a variation in the sensitivity to effects of pollutant exposure. This variation in sensitivity is linked to tolerance/resistance originating from pollutant-driven genetic selection (Ribeiro and Lopes, 2013). Indeed metal directional selection in plants has been studied for several years (Babst-Kostecka et al., 2016; Meyer et al., 2010).

6.2. Biomarkers Interest in Water Quality Bio-Monitoring

All the references of this review clearly highlight the usefulness of biomarkers to monitor environmental quality. The biomarker use must be validated and the results compared to those obtained by other methods. In an ecosystem, biotic (bacteria, viruses, parasites, competition, etc.) and abiotic factors (ETM, metalloids, pesticides, temperature, pH, hydrodynamic, etc.) affect organisms. Thus, before selecting the biomarker, it is necessary to know the xenobiotic impacts on organism metabolism (reproduction, immunity, assimilation...) and before experimental studies in laboratory or *in situ,* a meticulous state of the art must be done. Studying xenobiotic toxicity as well as their long-time effect in the environment could be done by biomarkers monitoring (Ferrat et al., 2003; Vasseur and Cossu-Leguille, 2003). Thus, environmental managers could early counteract the adverse effects of these contaminants before ecosystem degradation.

6.2.1. Potential Macrophyte Biomarkers

Biomarkers provide a dynamic and powerful approach to understand the effect of contaminants on organisms. In macrophytes, several types of biomarkers can be distinguished and studied:

- Biochemical markers such as hydrogen peroxide concentration, PCs and PAs concentrations, chlorophyll pigment and MDA content, enzymes of the antioxidant defense system like APX, CAT, SOD. As previously said in this chapter, arsenic induces oxidative stress leading to hydrogen peroxide and ROS production (Delmail and Labrousse, 2012). Then, ROS lead themselves to lipoperoxydation in chloroplasts and consequently to a degradation of chlorophyll pigments (Li et al., 2006; Rahman et al., 2007; Stoeva and Bineva, 2003) as well as an increase in MDA concentration. In order to reduce the cellular impacts of this metalloid, metal/metalloids-complexing molecules can be produced by the plant in order to inactivate it by PC-assisted vacuolar sequestration (Pal and Rai, 2010). An increased activity of enzymes of the antioxidant defense system can also be observed.
- Physiological markers such as respiratory and photosynthetic intensities, and osmotic potential. It was demonstrated that photosynthetic pigments were affected by metallic contamination. Consequently, photosynthetic and respiratory intensities are also affected and thus could constitute biomarkers for As pollution. The osmotic potential is also affected by metal/metalloid pollution (Delmail, 2011; Krayem, 2015).
- Morphological markers. These biomarkers are the easiest and cheapest parameter for the determination of a possible phytotoxic effect of metals and metalloids (Jiang et al., 2000). Several parameters can be studied such as stem length, root length, shoot branching, biomass... (Krayem et al., 2016; Delmail, 2011; Mench and Bes, 2009).
- Histological biomarkers such as the diameter of xylem vessels and the establishment of new structures like double endodermis during copper stress (Delmail, 2011; Delmail et al., 2011). Some of these biomarkers, such as micronuclei are used for the evaluation of genotoxic effects of xenobiotics (Pourrut et al., 2011).

Table 3. Potential biomarkers for arsenic pollution in aquatic and terrestrial plants

| | Ref. | Plant model | Dose and Duration of As Exposure | Biomarkers | | | | |
				Morphological	Physiological	Biochemical	Histological/ Molecular
Lichen	Pisani et al., 2011	*Xanthoria parietina*	Arsenite (0.133; 1.333;13.33; 133.3 μM), 24 hours		Photosynthetic efficiency	H_2O_2 content, membrane lipid peroxidation, pigments content	
	Krayem, 2015; Krayem et al., 2016b	*Myriophyllum alterniflorum*	Arsenate (1.33 μM), 21-30 days	Roots length, main stem length, and ramifications length	Photosynthetic and respiratory activities	H_2O_2 content, membrane lipid peroxidation, pigments content, phytochelatines	Micronucleii formation
Aquatic plant	Duman et al., 2010	*Lemna minor*	Arsenite and arsenate (1, 4, 16 and 64 μM), 6 days			Membrane lipid peroxidation, pigments content, antioxidant enzymes, protein content	

Table 3. (Continued)

| | Ref. | Plant model | Dose and Duration of As Exposure | Biomarkers | | | | |
				Morphological	Physiological	Biochemical	Histological/ Molecular
Aquatic plant	Ozturk et al., 2010	*Nasturtium officinale*	Arsenite 1, 3, 5, 10, and 50 µM, 7 days	Growth rate		Membrane lipid peroxidation, pigments content, antioxidant enzymes, proline content	DNA damage
	Ozturk et al., 2010	*Nasturtium officinale*	Arsenite 1, 3, 5, 10, and 50 µM, 7 days	Growth rate		Membrane lipid peroxidation, pigments content, antioxidant enzymes, proline content	DNA damage
	Chen et al., 2015	*Vallisneria natans*	Arsenate (66.667 µM), 7 days			Chlorophyll content, malondialdehyde and thiol levels	

| | Ref. | Plant model | Dose and Duration of As Exposure | Biomarkers | | | |
				Morphological	Physiological	Biochemical	Histological/ Molecular
	Díaz et al., 2013	*Fontinalis antipyretica*	Arsenate (0.00133 µM to 133.33 µM), 22 days		The net photosynthesis and dark respiration rates		
Aquatic plant	Srivastava et al., 2013	*Hydrilla verticillata*	Arsenate 43µM during 96h		net photosynthetic rate, stomatal conductance, transpiration rate, electron transport rate, and substomatal CO_2 concentration	The rate of superoxide radical, reduced ascorbate content, pigments content	

Table 3. (Continued)

Ref.	Plant model	Dose and Duration of As Exposure	Biomarkers			
			Morphological	Physiological	Biochemical	Histological/ Molecular
Rofkar et al., 2014	*Azolla caroliniana and Lemna minor*	20 μM of Na₂HAsO₄, 14 days	Relative growth rate		Chlorophyll and Anthocyanin Content	
Iriel et al., 2015	*Vallisneria gigantea, Azolla filiculoides and Lemna minor*	Arsenate (264 μM) 5 and 10 days		Photosystem system II quantum yield, photosynthesis	Carotenoid and flavonoid contents	
Gross et al., 2016	*Myriophyllum spicatum*	0, 1.44, 7.2 and 36 mg As (V).kg⁻¹ dry sediment, 14 days	Growth, biomass allocation (leaf, stem and root mass fractions)		Pigments content	
Mishra et al., 2014	*Ceratophyllum demersum*	Arsenate (25 μM), 4 weeks		Photosynthesis	Pigments content, starch content, oxidative stress	
Mishra et al., 2016	*Ceratophyllum demersum*	Arsenate (0.5 μM), 2 weeks			Chlorophyll precursors, pigments content	

| | | | Biomarkers | | | |
Ref.	Plant model	Dose and Duration of As Exposure	Morphological	Physiological	Biochemical	Histological/ Molecular
Zhang et al., 2012	*Wolffia globosa*	Arsenate (0, 1, 5, 10, 30, 50, 100 mM), 5 d			Thiols content (phytochelatins and glutathione)	
Tripathi et al., 2014	*Najas indica*	arsenate (10–250 µM) and arsenite (1–50 µM), 7 days		Electric conductivity	Membrane lipid peroxidation, H_2O_2 content, protein content, thiols content, antioxidant enzymes, phytochelatin and phytochelatin synthase, amino acid profile	Tripathi et al., 2014)

Table 3. (Continued)

| | Ref. | Plant model | Dose and Duration of As Exposure | Biomarkers | | | |
				Morphological	Physiological	Biochemical	Histological/ Molecular
Terrestrial plants	Vromman et al., 2011	*Atriplex atacamensis*	Arsenate (0, 100 or 1000 μM), 14 and 28 days	Growth biomass	Water content	Polyamines content, glycinebetaine and non-protein thiol content	
	Mascher et al., 2002	*Trifolium pretense*	Arsenate (5, 10, 50 mg.Kg⁻¹ soil), 10 weeks			Antioxidant enzymes, pigments content, free polyamines	
	Stoeva et al., 2005	*Phaseolus vulgaris*	0, 2, 5 mg arsenate /dm⁻³ of soil, 5 days	Shoot length, root length, fresh mass	Leaf water potential, relative water content, transpiration rate, stomatal conductance, photosynthesis	Pigments content, lipid peroxidation, peroxidase activity and soluble protein content.	

Ref.	Plant model	Dose and Duration of As Exposure	Biomarkers			
			Morphological	Physiological	Biochemical	Histological/Molecular
Singh et al., 2006	*Pteris ensiformis; Pteris vittata*	Arsenate (0, 133 or 267 µM), 10 days			Total soluble protein content, pigments content, membrane lipid peroxidation, H_2O_2 content, ascorbate and glutathione pool	
Gusman et al., 2013	*Lactuca sativa*	Arsenate (52.8 mM), 3 days	Relative growth rate	Photosynthetic efficiency	Antioxidant enzymes	

Thus, these biomarkers are useful for the monitoring of environmental changes in aquatic ecosystems. The most pertinent and potential biomarkers in terrestrial and aquatic plants during As stress are summarized in Table 3.

CONCLUSION

As a conclusion, this review emphasizes the interest of using aquatic plant biomarkers in biomonitoring of As contamination. In fact, biomarkers provide interesting information on the toxic effects of xenobiotics and are valuable tools to understand their mode of action (Brain and Cedergreen, 2009). Biomonitoring gives valuable information about the alterations occurring in the ecosystem due to environmental stress. Many previous research studies have made great progress in the biomonitoring of metal pollution in aquatic systems. In these studies, a number of bioindicators including various species are proposed and can be used to detect metal occurrence in rivers. The use of aquatic macrophytes as bioindicators has some advantages such as tolerance to metal pollution, easy sampling and realization of laboratory studies. The main effects observed in aquatic plants exposed to metal/metalloid stress included biochemical (pigment content, ROS and associated molecules…), physiological (photosynthesis and respiration inhibition) and morphological alterations (growth inhibition, vascular vessel number decrease, etc). Metal/metalloids accumulation in plants is also investigated in order to evaluate their potential for bioremediation (Zhou et al., 2008). To validate the use of biomarkers in bioindication of water quality, more research is needed in a large range of natural and human-impacted environments. More than one specific biomarker is required to assess the status of complex system (Vasseur and Cossu-Leguille, 2003). Arsenic is a very toxic element for plants, animals and humans, and must be carefully studied as its accumulation along the food chain is of great concern in many countries (Hettick et al., 2015). Physiological and cellular mechanisms of As uptake and transport in plants are investigated.

However, the molecular basis of As accumulation in plants is still unknown. The role of PCs and PAs biomarkers needs also to be elucidated as they probably play a very important role in As-tolerance in plants. Therefore, future research should focus on molecular and biotechnological approaches like omics studies (genomics, transcriptomics, metabolomics, proteomics…; Kumar et al., 2015) in order to further investigate the mechanisms of plant responses to metal and metalloids stress (Hasanuzzaman et al., 2015).

ACKNOWLEDGMENTS

We wish to express our sincere thanks to Stephen Midgley for his editing and critical reading of the manuscript.

REFERENCES

Adriano, D.C., 2001. Bioavailability of Trace Metals, in: *Trace Elements in Terrestrial Environments.* Springer New York, pp. 61–89.

Aguiar, F.C., Segurado, P., Urbanič, G., Cambra, J., Chauvin, C., Ciadamidaro, S., Dörflinger, G., Ferreira, J., Germ, M., Manolaki, P., Minciardi, M.R., Munné, A., Papastergiadou, E., Ferreira, M.T., 2014. Comparability of river quality assessment using macrophytes: A multi-step procedure to overcome biogeographical differences. *Sci. Total Environ.* 476–477, 757–767. doi:10.1016/j.scitotenv.2013.10.021.

Aguirre, G., Pilon, M., 2016. Copper Delivery to Chloroplast Proteins and its Regulation. *Front. Plant Sci.* 6. doi:10.3389/fpls.2015.01250.

Ahsan, N., Lee, D.G., Alam, I., Kim, P.J., Lee, J.J., Ahn, Y.O., Kwak, S.S., Lee, I.J., Bahk, J.D., Kang, K.Y., Renaut, J., Komatsu, S., Lee, B.H., 2008. Comparative proteomic study of arsenic-induced differentially expressed proteins in rice roots reveals glutathione plays a central role during As stress. *Proteomics* 8, 3561–3576. doi:10.1002/pmic.2007 01189.

Amiard-Triquet, C., Rainbow, P.S., 2009. *Environmental Assessment of Estuarine Ecosystems: A Case Study*. CRC Press.

Anjum, N.A., Ahmad, I., Mohmood, I., Pacheco, M., Duarte, A.C., Pereira, E., Umar, S., Ahmad, A., Khan, N.A., Iqbal, M., Prasad, M.N.V., 2012. Modulation of glutathione and its related enzymes in plants' responses to toxic metals and metalloids—A review. *Environ. Exp. Bot.* 75, 307–324. doi:10.1016/j.envexpbot.2011.07.002.

Ansari, A.A., Gill, S.S., Abbas, Z.K., Naeem, M., 2016. *Plant Biodiversity: Monitoring, Assessment and Conservation*. CABI.

Ardestani, M.M., van Straalen, N.M., van Gestel, C.A.M., 2014. The relationship between metal toxicity and biotic ligand binding affinities in aquatic and soil organisms: A review. *Environ. Pollut.* 195, 133–147. doi:10.1016/j.envpol.2014.08.020.

Arduini, I., Godbold, D.L., Onnis, A., 1995. Influence of copper on root growth and morphology of *Pinus pinea* L. and *Pinus pinaster* Ait. seedlings. *Tree Physiol.* 15, 411–415. doi:10.1093/treephys/15.6.411.

Augustynowicz, J., Gajewski, Z., Kostecka-Gugała, A., Wróbel, P., Kołton, A., 2016. Accumulation patterns of Cr in *Callitriche* organs—qualitative and quantitative analysis. *Environ. Sci. Pollut. Res.* 23, 2669–2676. doi:10.1007/s11356-015-5499-y.

Babst-Kostecka, A.A., Waldmann, P., Frérot, H., Vollenweider, P., 2016. Plant adaptation to metal polluted environments—Physiological, morphological, and evolutionary insights from *Biscutella laevigata*. *Environ. Exp. Bot.* 127, 1–13. doi:10.1016/j.envexpbot.2016.03.001.

Baccouch, S., Chaoui, A., Ferjani, E.E., 1998. Nickel-induced oxidative damage and antioxidant responses in *Zea mays* shoots. *Plant Physiol. Biochem.* 36, 689–694. doi:10.1016/S0981-9428(98)80018-1.

Baize, D., Sterckeman, T., 2001. Of the necessity of knowledge of the natural pedo-geochemical background content in the evaluation of the contamination of soils by trace elements. *Sci. Total Environ.* 264, 127–139. doi:10.1016/S0048-9697(00)00615-X.

Balestrasse, K.B., Gallego, S.M., Benavides, M.P., Tomaro, M.L., 2005. Polyamines and proline are affected by cadmium stress in nodules and roots of soybean plants. *Plant Soil* 270, 343–353. doi:10.1007/s11104-004-1792-0.

Barchowsky, A., Dudek, E.J., Treadwell, M.D., Wetterhahn, K.E., 1996. Arsenic induces oxidant stress and NF-KB activation in cultured aortic endothelial cells. *Free Radic. Biol. Med.* 21, 783–790. doi:10.1016/0891-5849(96)00174-8.

Béraud, E., Cotelle, S., Leroy, P., Férard, J.F., 2007. Genotoxic effects and induction of phytochelatins in the presence of cadmium in *Vicia faba* roots. *Mutat. Res. Toxicol. Environ. Mutagen.* 633, 112–116. doi:10.1016/j.mrgentox.2006.05.013.

Bertoldi, D., Tassoni, A., Martinelli, L., Bagni, N., 2004. Polyamines and somatic embryogenesis in two *Vitis vinifera* cultivars. *Physiol. Plant.* 120, 657–666. doi:10.1111/j.0031-9317.2004.0282.x.

Besse, J.P., Geffard, O., Coquery, M., 2012. Relevance and applicability of active biomonitoring in continental waters under the Water Framework Directive. *TrAC Trends Anal. Chem.*, Chemical Monitoring Activity for the Implementation of the Water Framework Directive 36, 113–127. doi:10.1016/j.trac.2012.04.004.

Birben, E., Sahiner, U.M., Sackesen, C., Erzurum, S., Kalayci, O., 2012. Oxidative Stress and Antioxidant Defense: *World Allergy Organ. J.* 5, 9–19. doi:10.1097/WOX.0b013e3182439613.

Böcük, H., Yakar, A., Türker, O.C., 2013. Assessment of *Lemna gibba* L. (duckweed) as a potential ecological indicator for contaminated aquatic ecosystem by boron mine effluent. *Ecol. Indic.* 29, 538–548. doi:10.1016/j.ecolind.2013.01.029.

Borboa, L., Torre, C. de L., 1996. The Genotoxicity of Zn(II) and Cd(II) in *Allium cepa* Root Meristematic Cells. *New Phytol.* 134, 481–486.

Bouchereau, A., Aziz, A., Larher, F., Martin-Tanguy, J., 1999. Polyamines and environmental challenges: recent development. *Plant Sci.* 140, 103–125. doi:10.1016/S0168-9452(98)00218-0.

Bousquet C., 1997. *Spéciation de l'arsenic en milieu marin*, rapport I.U.P. ENTES, Aix Marseille II, 1997 [Speciation of arsenic in Marin water, report from I.U.P. ENTES, Aix Marseille II, 1997].

Brain, R.A., Cedergreen, N., 2009. Biomarkers in Aquatic Plants: Selection and Utility, in: *Reviews of Environmental Contamination and Toxicology* Volume 198. Springer New York, New York, NY, pp. 1–61.

Buffle, J., DeVitre, R.R., 1993. *Chemical and Biological Regulation of Aquatic Systems.* CRC Press.

Buzier, R., Charriau, A., Corona, D., Lenain, J.F., Fondanèche, P., Joussein, E., Poulier, G., Lissalde, S., Mazzella, N., Guibaud, G., 2014. DGT-labile As, Cd, Cu and Ni monitoring in freshwater: Toward a framework for interpretation of in situ deployment. *Environ. Pollut.* 192, 52–58. doi:10.1016/j.envpol.2014.05.017.

Calow, P.P., 2009. Handbook of Ecotoxicology. John Wiley & Sons.

Cao, T., Xie, P., Ni, L., Wu, A., Zhang, M., Wu, S., Smolders, A.J.P., 2007. The role of NH_4^+ toxicity in the decline of the submersed macrophyte *Vallisneria natans* in lakes of the Yangtze River basin, China. *Mar. Freshw. Res.* 58, 581. doi:10.1071/MF06090.

Champeau, olivier, 2005. Biomarqueurs d'effets chez *C. Fluminea* : du développement en laboratoire à l'application en mésocosme [Biomarkers of effect in *C.Fluminea*: a study from the laboratory to a mesocosm application] [WWW Document]. URL http://www.theses. fr/09494914X (accessed 12.19.16).

Chandrakar, V., Naithani, S.C., Keshavkant, S., 2016. Arsenic-induced metabolic disturbances and their mitigation mechanisms in crop plants: A review. *Biologia* (Bratisl.) 71, 367–377. doi:10.1515/biolog-2016-0052.

Chaoui, A., Mazhoudi, S., Ghorbal, M.H., El Ferjani, E., 1997. Cadmium and zinc induction of lipid peroxidation and effects on antioxidant enzyme activities in bean (*Phaseolus vulgaris* L.). *Plant Sci.* 127, 139–147. doi:10.1016/S0168-9452(97)00115-5.

Chatenet, P., Froissard, D., Cook-Moreau, J., Hourdin, P., Ghestem, A., Botineau, M., Haury, J., 2006. Populations of *Myriophyllum alterniflorum* L. as bioindicators of pollution in acidic to neutral rivers in the Limousin region. *Hydrobiologia* 570, 61–65. doi:10.1007/s10750-006-0162-8.

Chen, G., Liu, X., Brookes, P.C., Xu, J., 2015. Opportunities for Phytoremediation and Bioindication of Arsenic Contaminated Water Using a Submerged Aquatic Plant: *Vallisneria natans* (lour.) *Hara. Int. J. Phytoremediation* 17, 249–255. doi:10.1080/15226514.2014.883496.

Chen, G., White, P.A., 2004. The mutagenic hazards of aquatic sediments: a review. *Mutat. Res. Mutat. Res.,* The Sources and Potential Hazards of Mutagens in Complex Environmental Matrices 567, 151–225. doi:10.1016/j.mrrev.2004.08.005.

Chugh, L.K., Sawhney, S.K., 1999. Photosynthetic activities of *Pisum sativum* seedlings grown in presence of cadmium. *Plant Physiol. Biochem.* 37, 297–303. doi:10.1016/S0981-9428(99)80028-X.

Clemens, S., 2001. Molecular mechanisms of plant metal tolerance and homeostasis. *Planta* 212, 475–486. doi:10.1007/s004250000458.

Clemens, S., Ma, J.F., 2016. Toxic Heavy Metal and Metalloid Accumulation in Crop Plants and Foods. *Annu. Rev. Plant Biol.* 67, 489–512. doi:10.1146/annurev-arplant-043015-112301.

Cobbett, C.S., 2000. Phytochelatins and Their Roles in Heavy Metal Detoxification. *Plant Physiol.* 123, 825–832. doi:10.1104/pp.123.3.825.

Conesa, H.M., Schulin, R., Nowack, B., 2010. Suitability of using diffusive gradients in thin films (DGT) to study metal bioavailability in mine tailings: possibilities and constraints. *Environ. Sci. Pollut. Res.* 17, 657–664. doi:10.1007/s11356-009-0254-x.

Cosio, C., Dunand, C., 2009. Specific functions of individual class III peroxidase genes. *J. Exp. Bot.* 60, 391–408. doi:10.1093/jxb/ern318.

Cullen, W.R., Reimer, K.J., 1989. Arsenic speciation in the environment. *Chem. Rev.* 89, 713–764. doi:10.1021/cr00094a002.

da Silva, E.C., de Albuquerque, M.B., Azevedo Neto, A.D. de, Silva Junior, C.D. da, 2013. Drought and Its Consequences to Plants – From

52   Maha Krayem, Véronique Deluchat, Raphaël Decou et al.

Individual to Ecosystem, in: Akinci, S. (Ed.), *Responses of Organisms to Water Stress. InTech.*

Daby, D., 2006. Coastal Pollution and Potential Biomonitors of Metals in Mauritius. *Water. Air. Soil Pollut.* 174, 63–91. doi:10.1007/s11270-005-9035-4.

Dazy, M., Béraud, E., Cotelle, S., Meux, E., Masfaraud, J.F., Férard, J.F., 2008. Antioxidant enzyme activities as affected by trivalent and hexavalent chromium species in *Fontinalis antipyretica* Hedw. *Chemosphere* 73, 281–290. doi:10.1016/j.chemosphere.2008.06.044.

Dazy, M., Masfaraud, J.F., Férard, J.F., 2009. Induction of oxidative stress biomarkers associated with heavy metal stress in *Fontinalis antipyretica* Hedw. *Chemosphere* 75, 297–302. doi:10.1016/j.chemosphere.2008.12.045.

Delmail, D., 2011. *Contribution de Myriophyllum alterniflorum et de son périphyton à la biosurveillance de la qualité des eaux face aux métaux lourds [Contribution of Myriophyllum alterniflorum and its periphyton to the biomonitoring of water quality versus metallic stress].* Limoges.

Delmail, D., Abasq, M.L., Courtel, P., Rouaud, I., Labrousse, P., 2013. DNA damage protection, antioxidant and free-radical scavenging activities of *Myriophyllum alterniflorum* DC (Haloragaceae) vegetative parts. *Acta Bot. Gallica* 160, 165–172. doi:10.1080/12538078.2013.799046.

Delmail, D., Labrousse, P., 2012. Plant Ageing, a Counteracting Agent to Xenobiotic Stress, in: Nagata, T. (Ed.), *Senescence.* InTech.

Delmail, D., Labrousse, P., Hourdin, P., Larcher, L., Moesch, C., Botineau, M., 2011. Differential responses of *Myriophyllum alterniflorum* DC (Haloragaceae) organs to copper: physiological and developmental approaches. *Hydrobiologia* 664, 95–105. doi:10.1007/s10750-010-0589-9.

Dhankher, O.P., Rosen, B.P., McKinney, E.C., Meagher, R.B., 2006. Hyperaccumulation of arsenic in the shoots of Arabidopsis silenced for arsenate reductase (ACR2). *Proc. Natl. Acad. Sci.* 103, 5413–5418. doi:10.1073/pnas.0509770102.

Di Toro, D.M., Allen, H.E., Bergman, H.L., Meyer, J.S., Paquin, P.R., Santore, R.C., 2001. Biotic ligand model of the acute toxicity of metals. 1. Technical Basis. *Environ. Toxicol. Chem.* 20, 2383–2396. doi:10.1002/etc.5620201034.

Díaz, S., Villares, R., Vázquez, M.D., Carballeira, A., 2013. Physiological Effects of Exposure to Arsenic, Mercury, Antimony and Selenium in the Aquatic Moss *Fontinalis antipyretica* Hedw. *Water. Air. Soil Pollut.* 224, 1659. doi:10.1007/s11270-013-1659-1.

Dočekalová, H., Škarpa, P., Dočekal, B., 2015. Diffusive gradient in thin films technique for assessment of cadmium and copper bioaccessibility to radish (Raphanus sativus). *Talanta* 134, 153–157. doi:10.1016/j.talanta.2014.11.014.

Duester, L., van der Geest, H.G., Moelleken, S., Hirner, A.V., Kueppers, K., 2011. Comparative phytotoxicity of methylated and inorganic arsenic- and antimony species to *Lemna minor, Wolffia arrhiza and Selenastrum capricornutum. Microchem. J., Antimony: Emerging Global Contaminant in the Environment* 97, 30–37. doi:10.1016/j.microc.2010.05.007.

Duffus, J.H., 2002. "Heavy metals" a meaningless term? (IUPAC Technical Report). *Pure Appl. Chem.* 74. doi:10.1351/pac 200274050793.

Duman, F., Ozturk, F., Aydin, Z., 2010. Biological responses of duckweed (*Lemna minor* L.) exposed to the inorganic arsenic species As(III) and As(V): effects of concentration and duration of exposure. *Ecotoxicology* 19, 983–993. doi:10.1007/s10646-010-0480-5.

Elias, M., Wellner, A., Goldin-Azulay, K., Chabriere, E., Vorholt, J.A., Erb, T.J., Tawfik, D.S., 2012. The molecular basis of phosphate discrimination in arsenate-rich environments. *Nature* 491, 134–137. doi:10.1038/nature11517.

Faivre-Rampant, O., Kevers, C., Dommes, J., Gaspar, T., 2000. The recalcitrance to rooting of the micropropagated shoots of the rac tobacco mutant: Implications of polyamines and of the polyamine metabolism. *Plant Physiol. Biochem.* 38, 441–448. doi:10.1016/S0981-9428(00)00768-3.

Farooq, M.A., Islam, F., Ali, B., Najeeb, U., Mao, B., Gill, R.A., Yan, G., Siddique, K.H.M., Zhou, W., 2016. Arsenic toxicity in plants: Cellular and molecular mechanisms of its transport and metabolism. *Environ. Exp. Bot.* 132, 42–52. doi:10.1016/j.envexpbot.2016.08.004.

Ferrat, L., Pergent-Martini, C., Roméo, M., 2003. Assessment of the use of biomarkers in aquatic plants for the evaluation of environmental quality: application to seagrasses. *Aquat. Toxicol.* 65, 187–204. doi:10.1016/S0166-445X(03)00133-4.

Filipiak-Szok, A., Kurzawa, M., Szłyk, E., 2015. Determination of toxic metals by ICP-MS in Asiatic and European medicinal plants and dietary supplements. *J. Trace Elem. Med. Biol.* 30, 54–58. doi:10.1016/j.jtemb.2014.10.008.

Finnegan, P.M., Chen, W., 2012. Arsenic Toxicity: The Effects on Plant Metabolism. *Front. Physiol.* 3. doi:10.3389/fphys.2012.00182.

Flemming, C.A., Trevors, J.T., 1989. Copper toxicity and chemistry in the environment: a review. Water. *Air. Soil Pollut.* 44, 143–158. doi:10.1007/BF00228784.

Fusconi, A., Repetto, O., Bona, E., Massa, N., Gallo, C., Dumas-Gaudot, E., Berta, G., 2006. Effects of cadmium on meristem activity and nucleus ploidy in roots of *Pisum sativum* L. cv. Frisson seedlings. *Environ. Exp. Bot.* 58, 253–260. doi:10.1016/j.envexpbot.2005.09.008.

Garg, N., Singla, P., 2011. Arsenic toxicity in crop plants: physiological effects and tolerance mechanisms. *Environ. Chem. Lett.* 9, 303–321. doi:10.1007/s10311-011-0313-7.

Gebel, T.W., 2001. Genotoxicity of arsenical compounds. *Int. J. Hyg. Environ. Health* 203, 249–262. doi:10.1078/S1438-4639(04)70036-X.

Gichner, T., Patková, Z., Száková, J., Demnerová, K., 2004. Cadmium induces DNA damage in tobacco roots, but no DNA damage, somatic mutations or homologous recombination in tobacco leaves. *Mutat. Res. Toxicol. Environ. Mutagen.* 559, 49–57. doi:10.1016/j.mrgentox.2003.12.008.

Gill, S.S., Tuteja, N., 2010. Reactive oxygen species and antioxidant machinery in abiotic stress tolerance in crop plants. *Plant Physiol. Biochem.* 48, 909–930. doi:10.1016/j.plaphy.2010.08.016.

Gonsebatt, M.E., Vega, L., Salazar, A.M., Montero, R., Guzmán, P., Blas, J., Del Razo, L.M., García-Vargas, G., Albores, A., Cebrián, M.E., Kelsh, M., Ostrosky-Wegman, P., 1997. Cytogenetic effects in human exposure to arsenic. *Mutat. Res. Mutat. Res.*, *Arsenic: A Paradoxical Human Carcinogen* 386, 219–228. doi:10.1016/S1383-5742(97)00009-4.

Grill, E., Löffler, S., Winnacker, E.L., Zenk, M.H., 1989. Phytochelatins, the heavy-metal-binding peptides of plants, are synthesized from glutathione by a specific γ-glutamylcysteine dipeptidyl transpeptidase (phytochelatin synthase). *Proc. Natl. Acad. Sci.* 86, 6838–6842.

Groppa, M.D., Benavides, M.P., 2007. Polyamines and abiotic stress: recent advances. *Amino Acids* 34, 35–45. doi:10.1007/s00726-007-0501-8.

Groppa, M.D., Benavides, M.P., Tomaro, M.L., 2003. Polyamine metabolism in sunflower and wheat leaf discs under cadmium or copper stress. *Plant Sci.* 164, 293–299. doi:10.1016/S0168-9452(02)00412-0.

Gross, E.M., Nuttens, A., Paroshin, D., Hussner, A., 2016. Sensitive response of sediment-grown *Myriophyllum spicatum* L. to arsenic pollution under different CO_2 availability. *Hydrobiologia.* doi:10.1007/s10750-016-2956-7.

Guan, X., Dong, H., Ma, J., Jiang, L., 2009. Removal of arsenic from water: Effects of competing anions on As(III) removal in $KMnO_4$–Fe(II) process. *Water Res.* 43, 3891–3899. doi:10.1016/j.watres.2009.06.008.

Gulz, P.A., Gupta, S.K., Schulin, R., 2005. Arsenic accumulation of common plants from contaminated soils. *Plant Soil* 272, 337–347. doi:10.1007/s11104-004-5960-z.

Gupta, D.K., Tohoyama, H., Joho, M., Inouhe, M., 2004. Changes in the levels of phytochelatins and related metal-binding peptides in chickpea seedlings exposed to arsenic and different heavy metal ions. *J. Plant Res.* 117, 253–256. doi:10.1007/s10265-004-0152-8.

Gür, N., Türker, O.C., Böcük, H., 2016. Toxicity assessment of boron (B) by *Lemna minor* L. and *Lemna gibba* L. and their possible use as model plants for ecological risk assessment of aquatic ecosystems with boron pollution. *Chemosphere* 157, 1–9. doi:10.1016/j.chemosphere. 2016.04.138.

Gusman, G.S., Oliveira, J.A., Farnese, F.S., Cambraia, J., 2013. Arsenate and arsenite: the toxic effects on photosynthesis and growth of lettuce plants. *Acta Physiol. Plant.* 35, 1201–1209. doi:10.1007/s11738-012-1159-8.

Ha, S.B., Smith, A.P., Howden, R., Dietrich, W.M., Bugg, S., O'Connell, M.J., Goldsbrough, P.B., Cobbett, C.S., 1999. Phytochelatin Synthase Genes from *Arabidopsis* and the Yeast *Schizosaccharomyces pombe*. *Plant Cell Online* 11, 1153–1163. doi:10.1105/tpc.11.6.1153.

Hall, J.L., 2002. Cellular mechanisms for heavy metal detoxification and tolerance. *J. Exp. Bot.* 53, 1–11. doi:10.1093/jexbot/53.366.1.

Hartwig, A., 1998. Carcinogenicity of metal compounds: possible role of DNA repair inhibition. *Toxicol. Lett.* 102–103, 235–239. doi:10.1016/S0378-4274(98)00312-9.

Hasanuzzaman, M., Nahar, K., Hakeem, K.R., Öztürk, M., Fujita, M., 2015. Arsenic Toxicity in Plants and Possible Remediation, in: *Soil Remediation and Plants. Elsevier*, pp. 433–501.

Hawa Bibi, M., Asaeda, T., Azim, E., 2010. Effects of Cd, Cr, and Zn on growth and metal accumulation in an aquatic macrophyte, *Nitella graciliformis*. *Chem. Ecol.* 26, 49–56. doi:10.1080/0275754090 3468110.

Hettick, B.E., Cañas-Carrell, J.E., French, A.D., Klein, D.M., 2015. Arsenic: A Review of the Element's Toxicity, Plant Interactions, and Potential Methods of Remediation. *J. Agric. Food Chem.* 63, 7097–7107. doi:10.1021/acs.jafc.5b02487.

Hopkin, S.P., 1989. *Ecophysiology of Metals in Terrestrial Invertebrates*. Springer.

Hopkins, W.G., 2003. *Physiologie végétale [Plant physiology]*. De Boeck Supérieur.

Huang, H., Wu, J.Y., Wu, J.H., 2006. Heavy Metal Monitoring Using Bivalved Shellfish from Zhejiang Coastal Waters, East China Sea.

Environ. Monit. Assess. 129, 315–320. doi:10.1007/s10661-006-9364-9.

Hussain, S.S., Ali, M., Ahmad, M., Siddique, K.H.M., 2011. Polyamines: Natural and engineered abiotic and biotic stress tolerance in plants. *Biotechnol. Adv.* 29, 300–311. doi:10.1016/j.biotechadv.2011.01.003.

IARC (Ed.), 2012. *IARC monographs on the evaluation of carcinogenic risks to humans, volume 100 C, arsenic, metals, fibres, and dusts* [this publication represents the views and expert opinions of an IARC Working Group on the Evaluation of Carcinogenic Risks to Humans, which met in Lyon, 17 - 24 March 2009]. IARC, Lyon.

INERIS, 2014. Données technico-économiques sur les substances chimiques en France: cuivre, composés et alliages, DRC-14-136881-02236A, 91 p. (version mars 2015) [Technical and economic data on chemicals in France: copper, compounds and alloys, DRC-14-136881-02236A, 91 p. (March 2015 version)] [*WWW Document*]. URL //rsde.ineris.fr/ouhttp://www.ineris.fr/substances/fr/(accessed12.19.16).

INERIS, 2010. INERIS - Santé, Expertise en toxicologie chronique. Institut National de l'Environnement Industriel et des Risques (2010) - Fiche de données toxicologiques, environnementales des substances chimiques: Arsenic. 21 p, DRC-09-103112-11453A, Version N°4 - 2010 [Health, Expertise in chronic toxicology. National Institute for the Industrial Environment and Risks (2010) - Toxicological Data Sheet, Environmental Data for Chemical Substances: Arsenic. 21 p, DRC-09-103112-11453A, Version No. 4 - 2010] [*WWW Document*]. URL http://www.ineris.fr/substances/fr/page/21 (accessed 12.23.16).

Inouhe, M., 2005. Phytochelatins. *Braz. J. Plant Physiol.* 17. doi:10.1590/S1677-04202005000100006.

Iriel, A., Dundas, G., Fernández Cirelli, A., Lagorio, M.G., 2015. Effect of arsenic on reflectance spectra and chlorophyll fluorescence of aquatic plants. *Chemosphere* 119, 697–703. doi:10.1016/j.chemosphere.2014.07.066.

Islam, E., Khan, M.T., Irem, S., 2015. Biochemical mechanisms of signaling: Perspectives in plants under arsenic stress. *Ecotoxicol. Environ. Saf.* 114, 126–133. doi:10.1016/j.ecoenv.2015.01.017.

Jiang, W., Liu, D., Li, H., 2000. Effects of Cu2+ on root growth, cell division, and nucleolus of *Helianthus annuus* L. *Sci. Total Environ.* 256, 59–65. doi:10.1016/S0048-9697(00)00470-8.

Kaiser, J., 2001. *Bioindicators and Biomarkers of Environmental Pollution and Risk Assessment.* Science Publishers.

Keller, M., 2015. Chapter 4 - Photosynthesis and Respiration, in: Keller, M. (Ed.), *The Science of Grapevines* (Second Edition). Academic Press, San Diego, pp. 125–143.

Kelter, P.B., Mosher, M.D., Scott, A., 2008. *Chemistry: The Practical Science.* Cengage Learning.

Königshofer, H., Lechner, S., 2002. Are polyamines involved in the synthesis of heat-shock proteins in cell suspension cultures of tobacco and alfalfa in response to high-temperature stress? *Plant Physiol. Biochem.* 40, 51–59. doi:10.1016/S0981-9428(01)01347-X.

Koo, B.J., Chen, W., Chang, A.C., Page, A.L., Granato, T.C., Dowdy, R.H., 2010. A root exudates based approach to assess the long-term phytoavailability of metals in biosolids-amended soils. *Environ. Pollut.* 158, 2582–2588. doi:10.1016/j.envpol.2010.05.018.

Krayem, M., 2015. Etude des effets de l'arsenic et du cuivre sur un macrophyte aquatique, *Myriophyllum alterniflorum* D.C. : évaluation des biomarqueurs pour la détection précoce de pollution [Study of the effects of arsenic and copper on an aquatic macrophyte, *Myriophyllum alterniflorum* D.C.: evaluation of biomarkers for the early detection of pollution.]. Limoges.

Krayem, Baydoun, M., Deluchat, V., Lenain, J.F., Kazpard, V., Labrousse, P., 2016a. Absorption and translocation of copper and arsenic in an aquatic macrophyte *Myriophyllum alterniflorum* DC. in oligotrophic and eutrophic conditions. *Environ. Sci. Pollut. Res.* 1–8. doi:10.1007/s11356-016-6289-x.

Krayem, Deluchat, V., Rabiet, M., Cleries, K., Lenain, J.F., Saad, Z., Kazpard, V., Labrousse, P., 2016b. Effect of arsenate As (V) on the biomarkers of *Myriophyllum alterniflorum* in oligotrophic and eutrophic conditions. *Chemosphere* 147, 131–137. doi:10.1016/j.chemosphere.2015.12.093.

Kronzucker, Britto, D.T., 2002. NH4 + toxicity in higher plants: a critical review. *Plant Physiol* 159, 567–584.

Kumar, A., Taylor, M., Altabella, T., Tiburcio, A.F., 1997. Recent advances in polyamine research. *Trends Plant Sci.* 2, 124–130. doi:10.1016/S1360-1385(97)01013-3.

Kumar, S., Dubey, R.S., Tripathi, R.D., Chakrabarty, D., Trivedi, P.K., 2015. Omics and biotechnology of arsenic stress and detoxification in plants: Current updates and prospective. *Environ. Int.* 74, 221–230. doi:10.1016/j.envint.2014.10.019.

Kusano, T., Yamaguchi, K., Berberich, T., Takahashi, Y., 2007. Advances in polyamine research in 2007. *J. Plant Res.* 120, 345–350. doi:10.1007/s10265-007-0074-3.

Lagadic, L., Caquet, Th, Amiard, JC, Ramade, F, 1998. *Utilisation de biomarqueurs pour la surveillance de la qualité de l'environnement [Use of biomarkers for monitoring the quality of the environment.].* Tec & Doc Lavoisier.

Lane, T.W., Saito, M.A., George, G.N., Pickering, I.J., Prince, R.C., Morel, F.M.M., 2005. Biochemistry: A cadmium enzyme from a marine diatom. *Nature* 435, 42–42. doi:10.1038/435042a.

Laperche, V., Bodénan, F., Dictor, M.C., Baranger, P., 2003. Guide méthodologique de l'arsenic, appliqué à la gestion des sites et sols pollués. BRGM/RP-52066-FR, 90 p., 5 Figure, 10 Table., 3 ann [Methodological guide to arsenic, applied to the management of polluted sites and soils. BRGM/RP-52066-EN, 90 p., 5 Figure, 10 Table, 3 ann] [*WWW Document*]. URL http://infoterre.brgm.fr/rapports/RP-52066-FR.pdf (accessed 12.23.16).

Li, C., Feng, S., Shao, Y., Jiang, L., Lu, X., Hou, X., 2007. Effects of arsenic on seed germination and physiological activities of wheat seedlings. *J. Environ. Sci.* 19, 725–732. doi:10.1016/S1001-0742(07)60121-1.

Li, N., Wang, J., Song, W.Y., 2016. Arsenic Uptake and Translocation in Plants. *Plant Cell Physiol.* 57, 4–13. doi:10.1093/pcp/pcv143.

Li, W.X., Chen, T.B., Huang, Z.C., Lei, M., Liao, X.Y., 2006. Effect of arsenic on chloroplast ultrastructure and calcium distribution in arsenic hyperaccumulator *Pteris vittata* L. *Chemosphere* 62, 803–809. doi:10.1016/j.chemosphere.2005.04.055.

Liehr, G.A., Zettler, M.L., Leipe, T., Witt, G., 2005. The ocean quahog *Arctica islandica* L.: a bioindicator for contaminated sediments. *Mar. Biol.* 147, 671–679. doi:10.1007/s00227-005-1612-y.

Lin, R., Wang, X., Luo, Y., Du, W., Guo, H., Yin, D., 2007. Effects of soil cadmium on growth, oxidative stress and antioxidant system in wheat seedlings (*Triticum aestivum* L.). *Chemosphere* 69, 89–98. doi:10.1016/j.chemosphere.2007.04.041.

Linnik, P.N., 2015. Arsenic in Natural Waters: Forms of Occurrence, Peculiarities of Migration, and Toxicity (a Review). *Hydrobiol. J.* 51. doi:10.1615/HydrobJ.v51.i6.100.

Liu, C.W., Chen, Y.Y., Kao, Y.H., Maji, S.K., 2014. Bioaccumulation and Translocation of Arsenic in the Ecosystem of the Guandu Wetland, Taiwan. *Wetlands* 34, 129–140. doi:10.1007/s13157-013-0491-0.

Liu, D., Jiang, W., Gao, X., 2003. Effects of Cadmium on Root Growth, Cell Division and Nucleoli in Root Tip Cells of Garlic. *Biol. Plant.* 47, 79–83. doi:10.1023/A:1027384932338.

Ma, J.F., Yamaji, N., Mitani, N., Xu, X.Y., Su, Y.H., McGrath, S.P., Zhao, F.J., 2008. Transporters of arsenite in rice and their role in arsenic accumulation in rice grain. *Proc. Natl. Acad. Sci.* 105, 9931–9935.

Ma, L.Q., Komar, K.M., Tu, C., Zhang, W., Cai, Y., Kennelley, E.D., 2001. A fern that hyperaccumulates arsenic. *Nature* 409, 579–579. doi:10.1038/35054664.

MacFarlane, G.R., 2003. Chlorophyll a Fluorescence as a Potential Biomarker of Zinc Stress in the Grey Mangrove, *Avicennia marina* (Forsk.) *Vierh. Bull. Environ. Contam. Toxicol.* 70, 0090–0096. doi:10.1007/s00128-002-0160-0.

Maitani, T., Kubota, H., Sato, K., Yamada, T., 1996. The Composition of Metals Bound to Class III Metallothionein (Phytochelatin and Its Desglycyl Peptide) Induced by Various Metals in Root Cultures of *Rubia tinctorum. Plant Physiol.* 110, 1145–1150. doi:10.1104/pp. 110.4.1145.

Mallick, S., Sinam, G., Sinha, S., 2011. Study on arsenate tolerant and sensitive cultivars of *Zea mays* L.: Differential detoxification mechanism and effect on nutrients status. *Ecotoxicol. Environ. Saf.* 74, 1316–1324. doi:10.1016/j.ecoenv.2011.02.012.

Mangabeira, P.A., Ferreira, A.S., Almeida, A.A.F. de, Fernandes, V.F., Lucena, E., Souza, V.L., Júnior, A.J. dos S., Oliveira, A.H., Grenier-Loustalot, M.F., Barbier, F., Silva, D.C., 2011. Compartmentalization and ultrastructural alterations induced by chromium in aquatic macrophytes. *Bio Metals* 24, 1017–1026. doi:10.1007/s10534-011-9459-9.

Martin-Tanguy, J., 2001. Metabolism and function of polyamines in plants: recent development (new approaches). *Plant Growth Regul.* 34, 135–148. doi:10.1023/A:1013343106574.

Mascher, R., Lippmann, B., Holzinger, S., Bergmann, H., 2002. Arsenate toxicity: effects on oxidative stress response molecules and enzymes in red clover plants. *Plant Sci.* 163, 961–969.

Mazej, Z., Germ, M., 2009. Trace element accumulation and distribution in four aquatic macrophytes. *Chemosphere* 74, 642–647. doi:10.1016/j.chemosphere.2008.10.019.

MEDD et Agences de l'eau, 2003. Système d'évaluation de la qualité de l'eau des cours d'eau SEQ-Eau, Grilles d'évaluation version 2,21 mars 2003 [Water quality assessment system for watercourses SEQ-Water, Evaluation grids version 2, 21 March 2003] [*WWW Document*]. URL http://www.observatoire-eau-bretagne.fr/Ressources-et-documentation/Documents-de-planification/Systeme-d-evaluation-de-la-qualite-de-l-eau-des-cours-d-eau-SEQ-Eau (accessed 12.23.16).

Meharg, A.A., Hartley-Whitaker, J., 2002. Arsenic uptake and metabolism in arsenic resistant and nonresistant plant species: Tansley review no. 133. *New Phytol.* 154, 29–43. doi:10.1046/j.1469-8137.2002.00363.x.

Meharg, A.A., Macnair, M.R., 1992. Suppression of the high affinity phosphate uptake system: a mechanism of arsenate tolerance in Holcus lanatus L. *J. Exp. Bot.* 43, 519–524.

Mench, M., Bes, C., 2009. Assessment of Ecotoxicity of Topsoils from a Wood Treatment Site*1. *Pedosphere* 19, 143–155. doi:10.1016/S1002-0160(09)60104-1.

Meneguetti, D.U. de O., da Silva, F.C., 2012. Adaptation of the Micronucleus Technique in *Allium Cepa*, For Mutagenicity Analysis of the Jamari River Valley, Western Amazon, Brazil. *J. Environ. Anal. Toxicol.* 2. doi:10.4172/2161-0525.1000127.

Meng, X., Bang, S., Korfiatis, G.P., 2000. Effects of silicate, sulfate, and carbonate on arsenic removal by ferric chloride. *Water Res.* 34, 1255–1261. doi:10.1016/S0043-1354(99)00272-9.

Meyer, C.L., Kostecka, A.A., Saumitou-Laprade, P., Créach, A., Castric, V., Pauwels, M., Frérot, H., 2010. Variability of zinc tolerance among and within populations of the pseudometallophyte species *Arabidopsis halleri* and possible role of directional selection. *New Phytol.* 185, 130–142. doi:10.1111/j.1469-8137.2009.03062.x.

Michon, J., Deluchat, V., Al Shukry, R., Dagot, C., Bollinger, J.C., 2007. Optimization of a GFAAS method for determination of total inorganic arsenic in drinking water. *Talanta* 71, 479–485. doi:10.1016/j.talanta.2006.06.016.

Mishra, S., Alfeld, M., Sobotka, R., Andresen, E., Falkenberg, G., Küpper, H., 2016. Analysis of sublethal arsenic toxicity to *Ceratophyllum demersum* : subcellular distribution of arsenic and inhibition of chlorophyll biosynthesis. *J. Exp. Bot.* 67, 4639–4646. doi:10.1093/jxb/erw238.

Mishra, S., Srivastava, S., Tripathi, R.D., Govindarajan, R., Kuriakose, S.V., Prasad, M.N.V., 2006. Phytochelatin synthesis and response of antioxidants during cadmium stress in *Bacopa monnieri* L. *Plant Physiol. Biochem.* 44, 25–37. doi:10.1016/j.plaphy.2006.01.007.

Mishra, S., Stärk, H.J., Küpper, H., 2014. A different sequence of events than previously reported leads to arsenic-induced damage in *Ceratophyllum demersum* L. *Metallomics* 6, 444. doi:10.1039/c3mt00317e.

Mkandawire, M., Dudel, E.G., 2005. Accumulation of arsenic in *Lemna gibba* L. (duckweed) in tailing waters of two abandoned uranium mining sites in Saxony, Germany. *Sci. Total Environ.* 336, 81–89. doi:10.1016/j.scitotenv.2004.06.002.

Mkandawire, M., Lyubun, Y.V., Kosterin, P.V., Dudel, E.G., 2004. Toxicity of arsenic species to *Lemna gibba* L. and the influence of phosphate on arsenic bioavailability. *Environ. Toxicol.* 19, 26–34. doi:10.1002/tox.10148.

Monnet, F., Bordas, F., Deluchat, V., Baudu, M., 2006. Toxicity of copper excess on the lichen *Dermatocarpon luridum*: Antioxidant enzyme activities. *Chemosphere* 65, 1806–1813. doi:10.1016/j.chemosphere.2006.04.022.

Monnet, F., Bordas, F., Deluchat, V., Chatenet, P., Botineau, M., Baudu, M., 2005. Use of the aquatic lichen as bioindicator of copper pollution: Accumulation and cellular distribution tests. *Environ. Pollut.* 138, 456–462. doi:10.1016/j.envpol.2005.04.019.

Muller, S.L., Huggett, D.B., J. H. Rodgers, J., 2001. Effects of Copper Sulfate on *Typha latifolia* Seed Germination and Early Seedling Growth in Aqueous and Sediment Exposures. *Arch. Environ. Contam. Toxicol.* 40, 192–197. doi:10.1007/s002440010163.

Navakoudis, E., Lütz, C., Langebartels, C., Lütz-Meindl, U., Kotzabasis, K., 2003. Ozone impact on the photosynthetic apparatus and the protective role of polyamines. *Biochim. Biophys. Acta BBA - Gen. Subj.* 1621, 160–169. doi:10.1016/S0304-4165(03)00056-4.

Ngayila, N., Basly, J.P., Lejeune, A.H., Botineau, M., Baudu, M., 2007. *Myriophyllum alterniflorum* DC., biomonitor of metal pollution and water quality. Sorption/accumulation capacities and photosynthetic pigments composition changes after copper and cadmium exposure. *Sci. Total Environ.* 373, 564–571. doi:10.1016/j.scitotenv.2006.11.038.

Nielsen, S.L., 1993. A comparison of aerial and submerged photosynthesis in some Danish amphibious plants. *Aquat. Bot.* 45, 27–40. doi:10.1016/0304-3770(93)90050-7.

Olszewska, J.P., Meharg, A.A., Heal, K.V., Carey, M., Gunn, I.D.M., Searle, K.R., Winfield, I.J., Spears, B.M., 2016. Assessing the Legacy of Red Mud Pollution in a Shallow Freshwater Lake: Arsenic

Accumulation and Speciation in Macrophytes. *Environ. Sci. Technol.* 50, 9044–9052. doi:10.1021/acs.est.6b00942.

Ozturk, F., Duman, F., Leblebici, Z., Temizgul, R., 2010. Arsenic accumulation and biological responses of watercress (*Nasturtium officinale* R. Br.) exposed to arsenite. *Environ. Exp. Bot.* 69, 167–174. doi:10.1016/j.envexpbot.2010.03.006.

Pal, R., Rai, J.P.N., 2010. Phytochelatins: Peptides Involved in Heavy Metal Detoxification. *Appl. Biochem. Biotechnol.* 160, 945–963. doi:10.1007/s12010-009-8565-4.

Pfeil, B.E., Schoefs, B., Spetea, C., 2014. Function and evolution of channels and transporters in photosynthetic membranes. *Cell. Mol. Life Sci.* 71, 979–998. doi:10.1007/s00018-013-1412-3.

Pinto, E., Sigaud-kutner, T.C.S., Leitao, M.A.S., Okamoto, O.K., Morse, D., Colepicolo, P., 2003. Heavy metal-induced oxidative stress in alga. *J. Phycol.* 39, 1008–1018. doi:10.1111/j.0022-3646.2003.02-193.x.

Pirintsos, S.A., Kotzabasis, K., Loppi, S., 2004. Polyamine Production in Lichens Under Metal Pollution Stress. *J. Atmospheric Chem.* 49, 303–315. doi:10.1007/s10874-004-1239-2.

Pisani, T., Munzi, S., Paoli, L., Bačkor, M., Loppi, S., 2011. Physiological effects of arsenic in the lichen *Xanthoria parietina* (L.) *Th. Fr. Chemosphere* 82, 963–969. doi:10.1016/j.chemosphere.2010.10.079.

Pollard, A.J., Reeves, R.D., Baker, A.J.M., 2014. Facultative hyperaccumulation of heavy metals and metalloids. *Plant Sci.* 217–218, 8–17. doi:10.1016/j.plantsci.2013.11.011.

Poschenrieder, C., Barceló, J., 2004. Water Relations in Heavy Metal Stressed Plants, in: Prasad, M.N.V. (Ed.), *Heavy Metal Stress in Plants.* Springer Berlin Heidelberg, pp. 249–270.

Pourrut, B., Shahid, M., Dumat, C., Winterton, P., Pinelli, E., 2011. Lead Uptake, Toxicity, and Detoxification in Plants, in: Whitacre, D.M. (Ed.), *Reviews of Environmental Contamination and Toxicology* Volume 213. Springer New York, New York, NY, pp. 113–136.

Prasad, M.N.V., 1998. Metal- biomolecule complexes in plants: Occurrence, functions, and applications. *Analusis* 26, 25–27. doi:10.1051/analusis:199826060025.

Punshon, T., Jackson, B.P., Meharg, A.A., Warczack, T., Scheckel, K., Guerinot, M.L., 2017. Understanding arsenic dynamics in agronomic systems to predict and prevent uptake by crop plants. *Sci. Total Environ.* 581–582, 209–220. doi:10.1016/j.scitotenv.2016.12.111.

Rahman, M.A., Hasegawa, H., Rahman, M.M., Rahman, M.A., Miah, M.A.M., 2007. Accumulation of arsenic in tissues of rice plant (*Oryza sativa* L.) and its distribution in fractions of rice grain. *Chemosphere* 69, 942–948. doi:10.1016/j.chemosphere.2007.05.044.

Rahman, M.A., Hasegawa, H., Ueda, K., Maki, T., Rahman, M.M., 2008. Arsenic uptake by aquatic macrophyte *Spirodela polyrhiza* L.: Interactions with phosphate and iron. *J. Hazard. Mater.* 160, 356–361. doi:10.1016/j.jhazmat.2008.03.022.

Rand, G.M., 1995. *Fundamentals of Aquatic Toxicology: Effects, Environmental Fate and Risk Assessment.* CRC Press.

Raskin, I., Kumar, P.N., Dushenkov, S., Salt, D.E., 1994. Bioconcentration of heavy metals by plants. *Curr. Opin. Biotechnol.* 5, 285–290. doi:10.1016/0958-1669(94)90030-2.

Rau, S., Miersch, J., Neumann, D., Weber, E., Krauss, G.J., 2007. Biochemical responses of the aquatic moss *Fontinalis antipyretica* to Cd, Cu, Pb and Zn determined by chlorophyll fluorescence and protein levels. *Environ. Exp. Bot.* 59, 299–306. doi:10.1016/j.envexpbot.2006.03.001.

Rea, P.A., Vatamaniuk, O.K., Rigden, D.J., 2004. Weeds, Worms, and More. Papain's Long-Lost Cousin, Phytochelatin Synthase. *Plant Physiol.* 136, 2463–2474. doi:10.1104/pp.104.048579.

Requejo, R., Tena, M., 2005. Proteome analysis of maize roots reveals that oxidative stress is a main contributing factor to plant arsenic toxicity. *Phytochemistry* 66, 1519–1528. doi:10.1016/j.phytochem.2005.05.003.

Ribeiro, R., Lopes, I., 2013. Contaminant driven genetic erosion and associated hypotheses on alleles loss, reduced population growth rate and increased susceptibility to future stressors: an essay. *Ecotoxicology* 22, 889–899. doi:10.1007/s10646-013-1070-0.

Ritchie, R.J., Mekjinda, N., 2016. Arsenic toxicity in the water weed *Wolffia arrhiza* measured using Pulse Amplitude Modulation

Fluorometry (PAM) measurements of photosynthesis. *Ecotoxicol. Environ. Saf.* 132, 178–185. doi:10.1016/j.ecoenv.2016.06.004.

RNO, 2006. RNO. *Surveillance du Milieu Marin.* Travaux du Réseau National d'Observation de la qualité du milieu marin. Edition 2006. [National network of observation of the marin water quality, edition 2006].

Robinson, B., Kim, N., Marchetti, M., Moni, C., Schroeter, L., van den Dijssel, C., Milne, G., Clothier, B., 2006. Arsenic hyperaccumulation by aquatic macrophytes in the Taupo Volcanic Zone, New Zealand. E*nviron. Exp. Bot.* 58, 206–215. doi:10.1016/j.envexpbot.2005.08.004.

Rofkar, J.R., Dwyer, D.F., Bobak, D.M., 2014. Uptake and Toxicity of Arsenic, Copper, and Silicon in *Azolla caroliniana* and *Lemna minor. Int. J. Phytoremediation* 16, 155–166. doi:10.1080/15226514. 2012.759534.

Rozentsvet, O.A., Nesterov, V.N., Sinyutina, N.F., 2012. The effect of copper ions on the lipid composition of subcellular membranes in *Hydrilla verticillata. Chemosphere* 89, 108–113. doi:10.1016/ j.chemosphere.2012.04.034.

Sandalio, L.M., Dalurzo, H.C., Gómez, M., Romero-Puertas, M.C., Río, L.A. del, 2001. Cadmium-induced changes in the growth and oxidative metabolism of pea plants. *J. Exp. Bot.* 52, 2115–2126. doi:10.1093/ jexbot/52.364.2115.

Sanders, J.G., Vermersch, P.S., 1982. Response of marine phytoplankton to low levels of arsenate. *J. Plankton Res.* 4, 881–893. doi:10.1093/ plankt/4.4.881.

Sanità di Toppi, L., Gabbrielli, R., 1999. Response to cadmium in higher plants. *Environ. Exp. Bot.* 41, 105–130. doi:10.1016/S0098-8472(98) 00058-6.

Santore, R.C., Di Toro, D.M., Paquin, P.R., Allen, H.E., Meyer, J.S., 2001. Biotic ligand model of the acute toxicity of metals. 2. Application to acute copper toxicity in freshwater fish and *Daphnia. Environ. Toxicol. Chem.* 20, 2397–2402. doi:10.1002/etc.5620201035.

Satofuka, H., Fukui, T., Takagi, M., Atomi, H., Imanaka, T., 2001. Metal-binding properties of phytochelatin-related peptides. *J. Inorg. Biochem.* 86, 595–602. doi:10.1016/S0162-0134(01)00223-9.

Schröder, P., Lyubenova, L., Huber, C., 2009. Do heavy metals and metalloids influence the detoxification of organic xenobiotics in plants? *Environ. Sci. Pollut. Res.* 16, 795–804. doi:10.1007/s11356-009-0168-7.

Seregin, I.V., Ivanov, V.B., 2001. Physiological Aspects of Cadmium and Lead Toxic Effects on Higher Plants. *Russ. J. Plant Physiol.* 48, 523–544. doi:10.1023/A:1016719901147.

Shaibur, M.R., Kitajima, N., Sugawara, R., Kondo, T., Huq, S.M.I., Kawai, S., 2006. Physiological and mineralogical properties of arsenic-induced chlorosis in rice seedlings grown hydroponically. *Soil Sci. Plant Nutr.* 52, 691–700. doi:10.1111/j.1747-0765.2006.00085.x.

Sharma, P., Jha, A.B., Dubey, R.S., Pessarakli, M., 2012. Reactive Oxygen Species, Oxidative Damage, and Antioxidative Defense Mechanism in Plants under Stressful Conditions. *J. Bot.* 2012, 1–26. doi:10.1155/2012/217037.

Sharma, S.S., 2006. The significance of amino acids and amino acid-derived molecules in plant responses and adaptation to heavy metal stress. *J. Exp. Bot.* 57, 711–726. doi:10.1093/jxb/erj073.

Sharma, Sohn, M., 2009. Aquatic arsenic: Toxicity, speciation, transformations, and remediation. *Environ. Int.* 35, 743–759. doi:10.1016/j.envint.2009.01.005.

Shen, H.J., Xie, Y.F., Li, R.T., 1994. Effects of acid stress on polyamine levels, ion efflux, protective enzymes and macromolecular synthesis in cereal leaves. *Plant Growth Regul.* 14, 1–5. doi:10.1007/BF00024134.

Sheppard, S.C., 1992. Summary of phytotoxic levels of soil arsenic. *Water. Air. Soil Pollut.* 64, 539–550. doi:10.1007/BF00483364.

Shri, M., Kumar, S., Chakrabarty, D., Trivedi, P.K., Mallick, S., Misra, P., Shukla, D., Mishra, S., Srivastava, S., Tripathi, R.D., Tuli, R., 2009. Effect of arsenic on growth, oxidative stress, and antioxidant system in rice seedlings. *Ecotoxicol. Environ. Saf.* 72, 1102–1110. doi:10.1016/j.ecoenv.2008.09.022.

Singh, N., Ma, L.Q., Srivastava, M., Rathinasabapathi, B., 2006. Metabolic adaptations to arsenic-induced oxidative stress in *Pteris vittata* L and *Pteris ensiformis* L. Plant Sci. 170, 274–282. doi:10.1016/ j.plantsci. 2005.08.013.

Singh, R., Singh, S., Parihar, P., Singh, V.P., Prasad, S.M., 2015. Arsenic contamination, consequences and remediation techniques: A review. *Ecotoxicol. Environ. Saf.* 112, 247–270. doi:10.1016/j.ecoenv. 2014.10.009.

Smedley, P.L., Kinniburgh, D.G., 2002. A review of the source, behaviour and distribution of arsenic in natural waters. *Appl. Geochem.* 17, 517–568. doi:10.1016/S0883-2927(02)00018-5.

Smith, K.S., Balistrieri, L.S., Todd, A.S., 2015. Using biotic ligand models to predict metal toxicity in mineralized systems. *Appl. Geochem., Environmental Geochemistry of Modern Mining* 57, 55–72. doi:10. 1016/j.apgeochem.2014.07.005.

Sobkowiak, R., Deckert, J., 2003. Cadmium-induced changes in growth and cell cycle gene expression in suspension-culture cells of soybean. *Plant Physiol. Biochem.* 41, 767–772. doi:10.1016/S0981-9428(03) 00101-3.

Souguir, D., Ferjani, E., Ledoigt, G., Goupil, P., 2008. Exposure of *Vicia faba* and *Pisum sativum* to copper-induced genotoxicity. *Protoplasma* 233, 203–207. doi:10.1007/s00709-008-0004-9.

Srivastava, M., Ma, L.Q., Santos, J.A.G., 2006. Three new arsenic hyperaccumulating ferns. *Sci. Total Environ.* 364, 24–31. doi:10.1016 /j.scitotenv.2005.11.002.

Srivastava, S., Srivastava, A.K., Singh, B., Suprasanna, P., D'souza, S.F., 2013. The effect of arsenic on pigment composition and photosynthesis in *Hydrilla verticillata*. *Biol. Plant.* 57, 385–389. doi:10.1007/s10535-012-0288-7.

Steffens, J.C., 1990. The Heavy Metal-Binding Peptides of Plants. *Annu. Rev. Plant Physiol. Plant Mol. Biol.* 41, 553–575. doi:10.1146/annurev.pp.41.060190.003005.

Stoeva, N., Berova, M., Zlatev, Z., 2005. Effect of arsenic on some physiological parameters in bean plants. *Biol. Plant.* 49, 293–296.

Stoeva, N., Bineva, T., 2003. Oxidative changes and photosynthesis in oat plants grown in As-contaminated soil. *Bulg J Plant Physiol* 29, 87–95.

Swain, G., Adhikari, S., Mohanty, P., 2014. Phytoremediation of Copper and Cadmium from Water Using Water Hyacinth, *Eichhornia Crassipes*. *Int. J. Agric. Sci. Technol.* 2, 1. doi:10.14355/ijast. 2014.0301.01.

Takahashi, T., Kakehi, J.I., 2010. Polyamines: ubiquitous polycations with unique roles in growth and stress responses. *Ann. Bot.* 105, 1–6. doi:10.1093/aob/mcp259.

Tandy, S., Mundus, S., Yngvesson, J., de Bang, T.C., Lombi, E., Schjoerring, J.K., Husted, S., 2011. The use of DGT for prediction of plant available copper, zinc and phosphorus in agricultural soils. *Plant Soil* 346, 167–180. doi:10.1007/s11104-011-0806-y.

Tangahu, B.V., Sheikh Abdullah, S.R., Basri, H., Idris, M., Anuar, N., Mukhlisin, M., 2011. A Review on Heavy Metals (As, Pb, and Hg) Uptake by Plants through Phytoremediation. *Int. J. Chem. Eng.* 2011, e939161. doi:10.1155/2011/939161.

Tarazona, J.V., Ramos-Peralonso, M.J., 2014. Ecotoxicology, in: *Encyclopedia of Toxicology*. Elsevier, pp. 276–280.

Tawfik, D.S., Viola, R.E., 2011. Arsenate Replacing Phosphate: Alternative Life Chemistries and Ion Promiscuity. *Biochemistry* (Mosc.) 50, 1128–1134. doi:10.1021/bi200002a.

Tiburcio, A.F., Altabella, T., Borrell, A., Masgrau, C., 1997. Polyamine metabolism and its regulation. *Physiol. Plant.* 100, 664–674. doi:10.1111/j.1399-3054.1997.tb03073.x.

Tripathi, R.D., Singh, R., Tripathi, P., Dwivedi, S., Chauhan, R., Adhikari, B., Trivedi, P.K., 2014. Arsenic accumulation and tolerance in rootless macrophyte *Najas indica* are mediated through antioxidants, amino acids and phytochelatins. *Aquat. Toxicol.* 157, 70–80. doi:10.1016/j.aquatox.2014.09.011.

Ünyayar, S., Çelik, A., Çekiç, F.Ö., Gözel, A., 2006. Cadmium-induced genotoxicity, cytotoxicity and lipid peroxidation in *Allium sativum* and *Vicia faba*. *Mutagenesis* 21, 77–81. doi:10.1093/mutage/gel001.

USEPA, 2001. *National Primary Drinking Water Regulation, Federal Register*, vol. 66, 2001, p. 6976 [WWW Document]. URL (accessed 12.19.16).

Valério, L., De Meyer, M., Penel, C., Dunand, C., 2004. Expression analysis of the *Arabidopsis* peroxidase multigenic family. *Phytochemistry* 65, 1331–1342. doi:10.1016/j.phytochem.2004.04.017.

Van Den Broeck, K., Vandecasteele, C., Geuns, J.M.C., 1997. Determination of Arsenic by Inductively Coupled Plasma Mass Spectrometry in Mung Bean Seedlings for use as a Bio-indicator of Arsenic Contamination. *J. Anal. At. Spectrom.* 12, 987–991. doi:10.1039/a701610g.

van der Ent, A., Baker, A.J.M., Reeves, R.D., Pollard, A.J., Schat, H., 2013. Hyperaccumulators of metal and metalloid trace elements: Facts and fiction. *Plant Soil* 362, 319–334. doi:10.1007/s11104-012-1287-3.

Vasseur, P., Cossu-Leguille, C., 2003. Biomarkers and community indices as complementary tools for environmental safety. *Environ. Int., Secotox* S.I. 28, 711–717. doi:10.1016/S0160-4120(02)00116-2.

Villaescusa, I., Bollinger, J.C., 2008. Arsenic in drinking water: sources, occurrence and health effects (a review). *Rev. Environ. Sci. Biotechnol.* 7, 307–323. doi:10.1007/s11157-008-9138-7.

Vromman, D., Flores-Bavestrello, A., Šlejkovec, Z., Lapaille, S., Teixeira-Cardoso, C., Briceño, M., Kumar, M., Martínez, J.P., Lutts, S., 2011. Arsenic accumulation and distribution in relation to young seedling growth in *Atriplex atacamensis Phil. Sci. Total Environ.* 412–413, 286–295. doi:10.1016/j.scitotenv.2011.09.085.

Wettlaufer, S.H., Osmeloski, J., Weinstein, L.H., 1991. Response of polyamines to heavy metal stress in oat seedlings. *Environ. Toxicol. Chem.* 10, 1083–1088. doi:10.1002/etc.5620100813.

Wood, C.M., 2011. An introduction to metals in fish physiology and toxicology: basic principles, in: *Fish Physiology*. Elsevier, pp. 1–51.

Wu, L., Yi, H., Yi, M., 2010. Assessment of arsenic toxicity using *Allium/Vicia* root tip micronucleus assays. *J. Hazard. Mater.* 176, 952–956. doi:10.1016/j.jhazmat.2009.11.132.

Xing, W., Huang, W., Liu, G., 2010. Effect of excess iron and copper on physiology of aquatic plant *Spirodela polyrrhiza* (L.) *Schleid. Environ. Toxicol.* 25, 103–112. doi:10.1002/tox.20480.

Xu, W.H., Liu, H., Ma, Q.F., Xiong, Z.T., 2007. Root Exudates, Rhizosphere Zn Fractions, and Zn Accumulation of Ryegrass at Different Soil Zn Levels1. *Pedosphere* 17, 389–396. doi:10.1016/S1002-0160(07)60047-2.

Xue, P., Li, G., Liu, W., Yan, C., 2010. Copper uptake and translocation in a submerged aquatic plant *Hydrilla verticillata* (L.f.) Royle. *Chemosphere* 81, 1098–1103. doi:10.1016/j.chemosphere.2010.09.023.

Xue, P., Yan, C., 2011. Arsenic accumulation and translocation in the submerged macrophyte *Hydrilla verticillata* (L.f.) Royle. *Chemosphere* 85, 1176–1181. doi:10.1016/j.chemosphere.2011.09.051.

Yabanli, M., Yozukmaz, A., Sel, F., 2014. Heavy Metal Accumulation In the Leaves, Stem and Root of the Invasive Submerged Macrophyte *Myriophyllum spicatum* L. (Haloragaceae): An Example of Kadın Creek (Mugla, Turkey). *Braz. Arch. Biol. Technol.* 57. doi:10.1590/S1516-8913201401962.

Yadav, S.K., 2010. Heavy metals toxicity in plants: An overview on the role of glutathione and phytochelatins in heavy metal stress tolerance of plants. *South Afr. J. Bot.* 76, 167–179. doi:10.1016/j.sajb.2009.10.007.

Yamanaka, K., Hayashi, H., Tachikawa, M., Kato, K., Hasegawa, A., Oku, N., Okada, S., 1997. Metabolic methylation is a possible genotoxicity-enhancing process of inorganic arsenics. *Mutat. Res. Toxicol. Environ. Mutagen.* 394, 95–101. doi:10.1016/S1383-5718(97)00130-7.

Yang, H., Shi, G., Wang, H., Xu, Q., 2010. Involvement of polyamines in adaptation of *Potamogeton crispus* L. to cadmium stress. *Aquat. Toxicol.* 100, 282–288. doi:10.1016/j.aquatox.2010.07.026.

Yi, H., Wu, L., Jiang, L., 2007. Genotoxicity of arsenic evaluated by Allium-root micronucleus assay. *Sci. Total Environ.* 383, 232–236. doi:10.1016/j.scitotenv.2007.05.015.

Yin, L., Zhou, Y., Fan, X., Lu, R., 2002. Induction of Phytochelatins in *Lemna aequinoctialis* in Response to Cadmium Exposure. *Bull. Environ. Contam. Toxicol.* 68, 561–568. doi:10.1007/s001280291.

Yıldız, M., Ciğerci, İ.H., Konuk, M., Fatih Fidan, A., Terzi, H., 2009. Determination of genotoxic effects of copper sulphate and cobalt chloride in *Allium cepa* root cells by chromosome aberration and comet assays. *Chemosphere* 75, 934–938. doi:10.1016/j.chemosphere. 2009.01.023.

Zaka, R., Chenal, C., Misset, M.T., 2002. Study of external low irradiation dose effects on induction of chromosome aberrations in *Pisum sativum* root tip meristem. *Mutat. Res. Toxicol. Environ. Mutagen.* 517, 87–99. doi:10.1016/S1383-5718(02)00056-6.

Zangi, R., Filella, M., 2012. Transport routes of metalloids into and out of the cell: A review of the current knowledge. *Chem. Biol. Interact.* 197, 47–57. doi:10.1016/j.cbi.2012.02.001.

Zenk, M.H., 1996. Heavy metal detoxification in higher plants - a review. *Gene* 179, 21–30. doi:10.1016/S0378-1119(96)00422-2.

Zhang, H., Zhao, F.J., Sun, B., Davison, W., Mcgrath, S.P., 2001. A New Method to Measure Effective Soil Solution Concentration Predicts Copper Availability to Plants. *Environ. Sci. Technol.* 35, 2602–2607. doi:10.1021/es000268q.

Zhang, X., Hu, Y., Liu, Y., Chen, B., 2011. Arsenic uptake, accumulation and phytofiltration by duckweed (*Spirodela polyrhiza* L.). *J. Environ. Sci.* 23, 601–606. doi:10.1016/S1001-0742(10)60454-8.

Zhang, X., Uroic, M.K., Xie, W.Y., Zhu, Y.G., Chen, B.D., McGrath, S.P., Feldmann, J., Zhao, F.J., 2012. Phytochelatins play a key role in arsenic accumulation and tolerance in the aquatic macrophyte *Wolffia globosa*. *Environ. Pollut., Chemicals Management and Environmental Assessment of Chemicals in China* 165, 18–24. doi:10.1016/j.envpol. 2012.02.009.

Zhao, F.J., Ma, J.F., Meharg, A.A., McGrath, S.P., 2009. Arsenic uptake and metabolism in plants: Tansley review. *New Phytol.* 181, 777–794. doi:10.1111/j.1469-8137.2008.02716.x.

Zhou, Q., Zhang, J., Fu, J., Shi, J., Jiang, G., 2008. Biomonitoring: An appealing tool for assessment of metal pollution in the aquatic ecosystem. *Anal. Chim. Acta* 606, 135–150. doi:10.1016/j.aca. 2007.11.018.

Zotina, T., Medvedeva, M., Trofimova, E., Alexandrova, Y., Dementyev, D., Bolsunovsky, A., 2015. Chromosomal abnormalities in roots of aquatic plant *Elodea canadensis* as a tool for testing genotoxicity of bottom sediments. *Ecotoxicol. Environ. Saf.* 122, 384–391. doi:10.1016/j.ecoenv.2015.08.021.

In: Arsenic
Editor: Ratko Knežević

ISBN: 978-1-53612-461-3
© 2017 Nova Science Publishers, Inc.

Chapter 2

THE UPTAKE OF ARSENIC IN FRUITS AND VEGETABLES: A CONCERN FOR HUMAN HEALTH

Shah Md Golam Gousul Azam[1], PhD,
Tushar C. Sarker[2], PhD and Sabrina Naz[1], PhD
[1]Department of Botany, University of Rajshahi, Rajshahi, Bangladesh
[2]Department of Agriculture, University of Naples Federico II,
Portici, Naples, Italy

ABSTRACT

Arsenic is a food chain contaminant. Arsenic contamination of fruits and vegetables is now being the additional source of this toxic element to mankind in South and South-east Asia. Current evidence shows that arsenic uptake in dietary plants is proportionally related to the presence of arsenic in soils and irrigation water where the major factors are flooding, arsenic forms, microorganism, organic matter, origin, and type of soil and plant. Excessive accumulation of arsenic in fruits and vegetables poses a potential health risk to the populations with higher consumption of fruits and vegetables, particularly to the local population of arsenic-affected area. To date, little attention has been paid to the risk of using arsenic-contaminated fruits and vegetables. In this perspective, our aim is to introduce the arsenic uptake in some of the common and popular fruits and vegetables that have been a pathway of risk to human health.

Keywords: arsenic, fruits, vegetables, health risk

1. INTRODUCTION

Current evidence indicates that long-term use of arsenic contaminated water for irrigation results elevated arsenic levels in soils and consequently in food crops (Williams et al., 2005). Intake of the foods (e.g., cereals, vegetables, fruits, spices) may result in an intake of significant levels of arsenic (Rahman and Hasegawa, 2011; Bergqvist et al., 2014; Azam et al., 2016b) and arsenic consumption, even at low levels, leads to carcinogenesis to human (Mandal and Suzuki, 2002).A large number of the total population are suffering from chronic arsenic diseases through crops and water contamination in South and South-east Asia and Latin America (Rahman and Hasegawa, 2011; Bundschuh et al., 2012).

The contamination of arsenic reported by anthropogenic activities, coal activities (geogenic coal excluding coal burning), geogenic, mining activities, petroleum activities, volcanogenic activities (Murcott, 2012). The accumulation of arsenic in plants are significantly associated with the presence of arsenic in soils and irrigation waters where the major factors are flooding, arsenic forms, microorganisms, organic matters, origin, and types of soil and plant (Azam et al., 2016a). Since fruits such as mango (Rahman et al., 2013; Saha and Zaman, 2013), lychee (Saha and Zaman, 2013) and vegetables such as spinach (Bergqvist et al., 2014), lettuce (Caporale et al., 2014) can accumulate significant level of arsenic, long-term ingestion of these arsenic-contaminated fruits and vegetables can be a cause of non-carcinogenic and carcinogenic diseases (Figure 1) (Nriagu et al., 2007; Cai et al., 2015; Jiang et al., 2015; Joseph et al., 2015; Rehman et al., 2016). On the other hand, consumption of fruits and vegetables is recommended as healthy diet which is associated with the reduced risk for several degenerative diseases including some cancers, cardiovascular disease, obesity, and all-cause mortality (Bazzano, 2005; Wang et al., 2014). With this in concern, fruits and vegetables as the exposure sources of arsenic must be considered more thoughtfully.

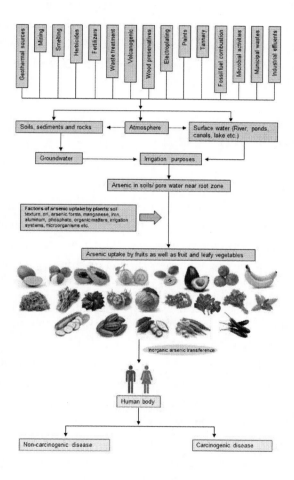

Figure 1. Arsenic accumulation by fruits as well as fruits and leafy vegetables from environment and its relation to human health risks. Arsenic releases into the environment by anthropogenic activities and natural processes such as mining, agricultural releases, industrial effluents, weathering of rocks and sediments, volcanic eruptions, geothermal activities, and wind-blown dust. Fruits and vegetables may uptake arsenic from arsenic-contaminated soil and/or water. As a consequence, arsenic easily can enter into the human body through consumption of fruits and vegetables as well as their products, and may cause of non-carcinogenic and carcinogenic disease.

Therefore, we highlight the soil and/or irrigation water contamination of arsenic in fruits and vegetables. Till to date, only a little attention has been paid to the risks of arsenic contaminated foods. In this review work, we have reviewed mainly on the existing literature related to the uptake of arsenic from the agricultural soil and irrigation water by fruits and vegetables. This compilation will be helpful for the scientists in working on pathways of human arsenic exposure that is a basis for human health risk assessment. Moreover, we present the health risk of arsenic to human by fruits and vegetables.

2. ARSENIC IN FRUITS AND VEGETABLES

Since the dietary plants can uptake arsenic (low level or more than the hygienic level), small differences in those foods nutritional composition will have major impacts on human health. In addition, consumers have a high demand for fruits and vegetables, which are the important dietary source of vitamins, phytochemicals, fibers and minerals (Hoffmann et al., 2003).

The accumulation of arsenic in the dietary plants is well known (Bundschuh et al., 2012; Azam et al., 2016b). The common and popular vegetablessuch as lettuce (Caporale et al., 2014), tomato (Pigna et al., 2012), cucumber (Hong et al., 2011a), and fruits such as mango (Rahman et al., 2013; Saha and Zaman, 2013), banana (Ahiamadjie et al., 2011) can accumulate high level of arsenic (Table 4). The transfer of arsenic from soils and/or water to the edible parts of plants is a key step in the way of arsenic entry into the human food chain. The uptake of arsenic occurred mainly in the root system and a lesser allocation in other parts of the plant such as stem, leave, grain and fruit (Azam et al., 2016b). Among others, leaves are more sensitive to arsenic contamination and can accumulate significantly. The accumulation of arsenic decreases the relative growth of leaves. A few studies reported that leafy vegetables such as spinach (Bhatti et al., 2013) water spinach (Li et al., 2014), taro (Nakwanit et al., 2011) which are grown in contaminated soils and/or water, accumulated high levels of arsenic. Similar findings has also been reported by many other authors, for instance, the arsenic accumulation ability of fruits and leafy vegetables in the contaminated soil and/or irrigated with arsenic-rich water (MacLean and Langille, 1981; Alam et al., 2003; Moyano et al., 2009; Li et al., 2014; Joseph et al., 2015).

2.1. Arsenic Contaminated Fruits

Baseline levels of arsenic in fruit part of the plant is low (10- 20%) comparing to other parts of the plants (Bohari et al., 2002; Bhatti et al., 2013; Bergqvist et al., 2014). Fruits can uptake significant amount of arsenic from arsenic-contaminated irrigation water and/or soil (Table 4).On the other hand, fruit is a good source of essential nutrients, dietary fiber, and beneficial phytochemicals, to improve health condition and decrease chronic disease risk (Bazzano, 2005).Due to the toxic effects of arsenic, long-term consumption of arsenic contaminated fruits is a concern (Mandal and Suzuki, 2002).

Themango (*Mangifera indica* L.), a juicy stone fruit, cultivated mostly for edible fruit, is a very popular fruit in the tropical region of the world. It is rich in carbohydrate, vitamin C, folate, and phytochemicals. Mango extracts could be used as nutraceuticals for the prophylaxis and treatment against oxidative stress (Table 1) (Ebeid et al., 2015; USDA, 2015).Till to date, there is no significant study on arsenic accumulation in Mango. However, the uptake and translocation of arsenic is occured in the fruits of mango plant. Though the accumulation of arsenic in fruits is low, it is a concern for its high popularity in the tropical region where people eat a good amount of fruit in the harvesting season as well as drink the juice of mango all the year round. The accumulation of arsenic in mango was 0.091 mg/kg grown in a contaminated area from west Bengal, India (Table 3) (Rahaman et al., 2013). In another study, the concentration of arsenic in the mango fruits ranged from 0.006 to 0.05 mg/kg with a mean value of 0.008 mg/kg (Liao et al., 2014). In addition, the concentration of arsenic was

0.128 mg/kg in mango sourced from market outlets of Rajshahi city, Bangladesh (Saha and Zaman, 2013).

Lychee (*Litchi chinensis* Sonn.), a popular fruit has a delicate, whitish pulp with a floral smell and a fragrant, sweet flavor. It is an important source of vitamin C, polyphenols (Table 1) which belong to anti-oxidant, anti-obesity, anti cancer activities (Aruoma et al., 2006; Qi et al., 2015; USDA, 2015). In Malda district of West Bengal, India, the concentration of arsenic in lychee fruit was 0.119 mg/kg (Rahaman et al., 2013) where the soil and groundwater were contaminated with arsenic (Table 3). In a market basket survey from Rajshahi, Bangladesh, the concentration of arsenic in lychee fruits was 0.125 mg/kg (Saha and Zaman, 2013).

Papaya (*Carica papaya* L.) is a very common fruit vegetable. People eat papaya as curry or fruits considering a very healthy food. It is a rich source of vitamin C, folate, carotenoids, polyphenols (Table 1) which have anticholinergic, antioxidant and antiproliferative activity (Gironés-Vilaplana et al., 2015; Rodriguez-Amaya, 2015; USDA, 2015). An investigation of arsenic in food composites, collected from the villagers, was carried out in arsenic-affected areas of the Murshidabad district, West Bengal. The accumulation of arsenic was 0.196 mg/kg and 0.373 mg/kg in two location of Domkal block, Murshidabad (Roychowdhury et al., 2002). In Samta village, Bangladesh, foodstuffs were irrigated with high arsenic-contaminated water (Table 3). The concentration of arsenic in papaya was 0.40 mg/kg (Alam et al., 2003). In west Bengal (India), the accumulation of arsenic in papaya varied from 0.123 to 0.276 mg/kg irrigated with arsenic-rich groundwater (Table 3) (Rahaman et al., 2013).

Guava (*Psidium guajava* L.) is a common tropical fruitcultivated in mainly tropical regions of the world. It is a potential source of dietary fiber and vitamin C, folic acid, carotenoids, and polyphenols (Table 1) which belongs to antioxidative, anti-inflammatory and antiproliferative activities (Gutiérrez et al., 2008; USDA, 2015). The concentration of arsenic in guava ranged from 0.028 to 0.056 mg/kg in a highly arsenic-contaminated groundwater area of west Bengal, India (Table 3) (Rahaman et al., 2013). In another study, the concentration of arsenic in guava fruits was 0.131

mg/kg collected from the market of Rajshahi city, Bangladesh (Saha and Zaman, 2013). The concentration of arsenic was measured in the leaf of guava from Zimapán mining areas, Mexico. The arsenic concentration of leaf was 8.4 mg/kg (Prieto-García et al., 2005).

Sapodilla (*Manilkara zapota* (L.) P Royan) fruit has an exceptionally sweet, malty flavor containing vitamin B6, vitamin C, riboflavin, niacin, vitamin E, manganese, potassium and dietary fiber (Table 1) (USDA, 2015). The concentration of As was reported in several cultivated plants from Zimapán mining areas, Mexico where the uptake of arsenic was found to be 2.8 mg/kg in sapodilla fruits (Prieto-García et al., 2005).

Avocado (*Persea Americana* Mill.) is commercially valuable and is cultivated in tropical and Mediterranean climates throughout the world. It contains B vitamins and vitamin K, vitamin C, vitamin E and potassium polysterols and carotenoids (Table 1). It can reduce cholesterol level and risk of metabolic syndrome (III et al., 2013; USDA, 2015; Wang et al., 2015).Very high arsenic accumulation (5.7 mg/kg) in avocado fruit was reported from Zimapán mining areas, Mexico (Prieto-García et al., 2005).

Ber (*Ziziphus mauritiana* Lam.) fruit is eaten raw, pickled or used in beverages containing vitamin C, amino acids, B vitamins, calcium, iron, and phosphorus (Table 1) (USDA, 2015). Ber was cultivated in an arsenic affected area of West Bengal, India. The uptake of arsenic was low (0.02 mg/kg) in ber fruits (Rahaman et al., 2013).

Banana (*Musa* × *paradisiaca* L.) is a staple source of starch for many tropical populations. Banana inflorescence (hearts) is considered as avegetable as well as curries, and fried foods in South and Southeast Asia. It is an important source of pro-vitamin A, carotenoids, phenolics, and amine compounds (Table 1). Antioxidant activity of banana promotes wound healing, and prevents depression (Pereira and Maraschin, 2015; USDA, 2015). The effect of gold mining activities on selected consumed food crops was evaluated. The concentration of arsenic in banana was significantly high (3.12–8.43 mg/kg) (Ahiamadjie et al., 2011) and in the Zimapán mining areas, Mexico, the accumulation of arsenic in banana fruits was also high (2.54 mg/kg) (Prieto-García et al., 2005).

2.2. Arsenic Contaminated Vegetables

The healthy effects of vegetable consumption are quite similar with fruits because of its nutrient composition, dietary fiber, and phytochemicals which promote health condition (Bazzano, 2005). Unfortunately, vegetable can uptake arsenic from the contaminated soil and/or water exceeding the maximum allowable limit (0.1 mg/kg) set by WHO/FAO (Rehman et al., 2016). All the vegetables reviewed in this paper were above the maximum allowable limit of arsenic accumulation in the edible part except basella (Table 4).

Lettuce (*Lactuca sativa* L.), an important dietary leafy vegetable is consumed fresh or in salad mixes. Due to its perception as healthier foods and a number of lettuce varieties have been reported to contain phenolic compounds, antioxidant activities and vitamin C (Table 2) (DuPont et al., 2000; Llorach et al., 2008; USDA, 2015).In nutrient solution, the concentration of arsenic in lettuce during *in vivo* bioavailability assays was 4.4 mg/kg (Juhasz et al., 2008). The concentration of arsenic in the roots was up to 3 folded higher than that observed in the leaves (2.10-2.66 mg/kg) (Caporale et al., 2014). It was noted that up to 32 mg As/kg was accumulated in lettuce cultivated in soil near CCA-treated poles (Cao and Ma, 2004) while 6.8 and 17.8 mg As/kg were noticed in lettuce cultivated in soil from an ex-industrial area (Warren et al., 2003).The concentration of arsenic in lettuce leaf (0.033 mg/kg) (Chang et al., 2014), (0.023 mg/kg) (Huang et al., 2006), (0.020-0.250) mg /kg (Matschullat, 2000), (0.022-0.035) mg/kg dry wt (Cobb et al., 2000) have been investigated and these were below the national and international limit (0.1 mg/kg) (McBride, 2013). The lower concentration of the metalloid in the leaves may suggest a lower uptake ability of the plant at a systemic level (Smith et al., 2009). In a field trial, the accumulation of arsenic in different vegetables was investigated in the vicinity of abandoned pyrite mines. The accumulation of arsenic in lettuce leaves varied from 0.2 to 1.8 mg/kg in a highly arsenic-contaminated soil (Alvarenga et al., 2014).A study was conducted in a historic peri-urban waste disposal sites near Manchester, UK to know the human health risk. The uptake of arsenic in lettuce was 0.055 mg/kg (Atkinson et al., 2012).

Spinach (*Spinacia oleracea* L.) is a popular leafy vegetable worldwide. It contains several active components, including flavonoids (Table 2), exhibit antioxidative, antiproliferative, and anti-inflammatory properties in biological systems. Spinach extracts have numerous healthy effects, such as chemo- and central nervous system protection, anticancer and anti-aging functions (Porrini et al., 2002; Lomnitski et al., 2003; USDA, 2015).The concentration of arsenic in root was high (83.10 mg As/kg) in glassworks (Gadderås), Sweden followed by shoot (7.0310 mg As/kg) (Bergqvist et al., 2014). In a previous study, the concentration of arsenic was 0.90 mg/kg that was irrigated with lake water in the area of Sindh, Pakistan (Arain et al., 2009). The concentration of arsenic in spinach leaves exceeded the Chinese maximum permissible concentration for inorganic arsenic (0.05 mg/kg) by a factor of 1.6 to 6.4 folded. Spinach presents a direct risk to human health where irrigation is in flooded condition with water having an arsenic concentration greater than 50 μg As/l. The uptake of arsenic increased in the order carrot < radish < tomato < spinach. Irrigation water with an arsenic concentration greater than 50 μg/l should be stopped for spinach cultivation where irrigation in flooded condition is practiced (Bhatti et al., 2013). In this case, Bangladesh has arsenic concentrations of groundwater from deep wells in 64 districts in the country and found that 43 had concentrations >50μg/l (Hossain, 2006).So, it is suggested that spinach cultivation should be dropped during the rainy season in those 43 districts of Bangladesh. In another study, highest arsenic value was 0.910 mg/kg in in spinach from west Bengal, India (Biswas et al., 2012). In addition, in a less contaminated soil and highly contaminated irrigation water (Table 3), the accumulation of arsenic in spinach ranged from 0.096 to 0.126 mg/kg in Bangladesh (Farid et al., 2003). In another study, the accumulation of arsenic and translocation of nutrient elements in Japanese Mustard spinach (JMS) grown hydroponically has been reported. The little concentration of arsenic had a stimulating effect on JMS fresh weight, thus, toxicity increased with increasing arsenic in the medium. In the shoot, phosphorus, potassium, calcium, magnesium, iron, manganese and zinc concentrations were limited (Shaibur and Kawai, 2010). In Taiwan, the uptake of arsenic by spinach was 0.046 mg/kg where the groundwater was contaminated (Kar et al., 2013). The concentration of arsenic in spinach varied from 0.13 to 0.18 mg/kg from the area adjacent to the Hazaribag tannery, Dhaka, Bangladesh (Islam et al., 2014).

Betel (*Piper betle* L.) leaf is mostly consumed in Asia and elsewhere in the world by some Asian emigrants, as betel quid or in paan. It is a very common social habit to chew in East and South Asian countries by 200 million populations daily (Norton, 1998). Betel leaf contains phenols, hydroxychavicol and chavibetol. It has antimutagenic and antioxidant activities (Paranjpe et al., 2013). Betel leaves were purchased from UK based ethnic shops in the cities of Leicester, Birmingham and London and the concentration of arsenic in betel leaves was 0.107 mg/kg (Al-Rmalli et al., 2011). In addition, the uptake of arsenic (0.094 mg/kg) was observed in piper betel leaves while it was observing as a commonly used medicinal herb in Ayurveda from west Bengal, India (Nema et al., 2014).

Cabbage (*Brassica oleracea* L.) is a leafy green or purple biennial plant, grown as an annual vegetable crop. It is a rich source of phytochemicals, vitamin K, vitamin C and dietary fibre which has antioxidant and cytoprotective activity (Table 2) (Kabir et al., 2015; USDA, 2015). In a field study, the accumulation of toxic elements by vegetables produced in the vicinity of abandoned pyrite mines was evaluated. The maximum total concentrations of arsenic in soil were extremely high (i.e., 1851 mg/kg) (Table 3). The accumulation of arsenic in cabbage ranged from 0.17 to 0.40 mg/kg (Alvarenga et al., 2014). In an arsenic-rich groundwater area of Southwestern Taiwan, the uptake arsenic in cabbage was 0.0126 mg/kg (Table 3) (Kar et al., 2013).Vegetables were cultivated in urban waste disposal near Manchester, UK to know the heavy metal contamination in foods. The accumulation of arsenic in cabbage was 0.085 mg/kg (Atkinson et al., 2012). In China, arsenic uptake in vegetables was investigated that was irrigated with wastewater. The accumulation of arsenic in cabbage was 0.28 mg/kg and the transfer factor ranged from 0.02

to 0.05 (Wang et al., 2012). In Nadia district of West Bengal, India, the accumulation of arsenic in cabbage was0.315 µg/kg (Samal et al., 2011). In Malda district of West Bengal, India, high levels of arsenic concentration in groundwater were observed (Table 3) and the concentration of arsenic in cabbage varied from 0.211 to 0.456 mg/kg (Rahaman et al., 2013).

Leaf of parsley (*Petroselinum crispum* (Mill.) Fuss) is often used as a garnish. Parsley is used in Middle Eastern, European, and Americal cooking. It is a source of protein, carbohydrate, dietary fibers, vitamins and minerals (Table 2) (USDA, 2015). It is an important source of flavonoid, apigenin, folic acid, vitamin, K, vitamin C and vitamin A which belongs to anti-inflammatory, antioxidant, antimicrobial, antiallergic and immunomodulatory activity (Kamboh et al., 2015). Parsley was cultivated in an urban waste disposal area near Manchester, United Kingdom. The accumulation of arsenic was 0.08 mg/kg (Atkinson et al., 2012). The level of arsenic measured in the cultivated plants from Zimapán mining areas, Mexico. In the leaf of parsley, the concentration of arsenic was 11 mg/kg followed by stem (4.4 mg/kg) (Prieto-García et al., 2005).

Peppermint (*Mentha* × *piperita* L.) is the oldest and most popular flavor of mint-flavoured confectionery and is often used in tea and for flavoring ice cream, chewing gum, and toothpaste. It is the major source of menthol, menthone and menthyl acetate. It has antioxidant, antibacterial and antispasmodic activity, and it is used for the treatment of irritable bowel syndrome (Khanna et al., 2014; Riachi and De Maria, 2015). In Sindh, Pakistan, foodstuffs were cultivated by arsenic-contaminated lake and canal water (Table 3). The concentration of arsenic in the leaves of peppermint was 1.20 and 0.45 mg/kg respectively (Arain et al., 2009). In another study, toxic metal concentrations were assessed in some spices and herbs. Among others, the concentration of arsenic was found to be 0.104 mg/kg (Karadaş and Kara, 2012). The uptake of arsenic by peppermint is grown in soil irrigated with canal water as control vegetable samples and soil irrigated with tube well water of three sub-district of Pakistan as tested vegetable samples. The accumulation of arsenic varied from 1.01 to 1.70 mg/kg (Baig and Kazi, 2012).

Table 1. Nutrient composition of different fruits

Type	Nutrient	Mango*	Lychee*	Papaya*	Guava*	Sapota*	Avocado*	Ber*	Banana*
Proximates	Water (g)	83.46	81.76	88.06	80.80	64.87	73.23	77.86	74.91
	Energy (kcal)	60	66	43	68	124	160	79	89
	Protein (g)	0.82	0.83	0.47	2.55	1.45	2.00	1.20	1.09
	Total lipid (g)	0.38	0.44	0.26	0.95	0.46	14.66	0.20	0.33
	Carbohydrate, by difference (g)	14.98	16.53	10.86	14.32	32.10	8.53	20.23	22.84
	Fiber, total dietary (g)	1.6	1.3	1.7	5.4	5.4	6.7	-	2.6
	Sugars, total (g)	13.66	15.23	7.82	8.92	20.14	0.66	-	12.23
Minerals	Calcium, Ca (mg)	11	5	20	18	18	12	21	5
	Iron, Fe (mg)	0.16	0.31	0.25	0.26	0.78	0.55	0.48	0.26
	Magnesium, Mg (mg)	10	10	21	22	11	29	10	27
	Phosphorus, P (mg)	14	31	10	40	26	52	23	22
	Potassium, K (mg)	168	171	182	417	454	485	250	358
	Sodium, Na (mg)	1	1	8	2	7	7	3	1
	Zinc, Zn (mg)	0.09	0.07	0.08	0.23	0.19	0.64	0.05	0.15
Vitamins	Vitamin C, total ascorbic acid (mg)	36.4	71.5	60.9	228.3	23.0	10.0	69.0	8.7
	Thiamin (mg)	0.028	0.011	0.023	0.067	0.013	0.067	0.020	0.031
	Riboflavin (mg)	0.038	0.065	0.027	0.040	0.116	0.130	0.040	0.073
	Niacin (mg)	0.669	0.603	0.357	1.084	1.432	1.738	0.900	0.665
	Vitamin B-6 (mg)	0.119	0.100	0.038	0.110	0.720	0.257	0.081	0.375
	Folate, DFE (µg)	43	14	37	49	0.00	81	-	20

Type	Nutrient	Mango*	Lychee*	Papaya*	Guava*	Sapota*	Avocado*	Ber*	Banana*
	Vitamin A, RAE (µg)	54	0.0	47	31	7	7	2	3
	Vitamin A, IU (IU)	1082	0.0	950	628	143	146	40	64
	Vitamin E (alpha-tocopherol) (mg)	0.90	0.07	0.30	0.73	2.11	2.07	-	0.10
	Vitamin K (phylloquinone) (µg)	4.2	0.4	2.6	2.6	0.00	21.0	-	0.5
Lipids	Fatty acids, total saturated (g)	0.092	0.099	0.081	0.272	0.169	2.126	-	0.112
	Fatty acids, total monounsatu-rated (g)	0.140	0.120	0.072	0.087	0.102	9.799	-	0.032
	Fatty acids, total polyunsaturated (g)	0.071	0.132	0.058	0.401	0.097	1.826	-	0.073

*Per 100 g fresh weight.

Adapted from the U.S. Department of Agriculture, Agriculture Research Service (USDA, 2015).

Table 2. Nutrient composition of different vegetables

Type	Nutrient	Lettuce*	Spinach*	Tomato*	Cucumber*	Bitter gourd*	Okra*	Brinjal/ Eggplant*	Amaranth*	Cabbage*	Parsley*	Peppermint*	Coriander*	Basella*
Proximates	Water (g)	94.98	91.40	93.00	96.73	94.03	89.58	92.30	91.69	92.18	87.71	78.65	92.21	93.10
	Energy (kcal)	15	23	23	12	17	33	25	23	25	36	70	23	19
	Protein (g)	1.36	2.86	1.20	0.59	1.00	1.93	0.98	2.46	1.28	2.97	3.75	2.13	1.80
	Total lipid (g)	0.15	0.39	0.20	0.16	0.17	0.19	0.18	0.33	0.10	0.79	0.94	0.52	0.30
	Carbohydrate, by difference (g)	2.87	3.63	5.10	2.16	3.70	7.45	5.88	4.02	5.80	6.33	14.89	3.67	3.40
	Fiber, total dietary (g)	1.3	2.2	1.1	0.7	2.8	3.2	3.0	-	2.5	3.3	8.0	2.8	-
	Sugars, total (g)	0.78	0.42	4.00	1.38	-	1.48	3.53	-	3.20	0.85	-	0.87	-
Minerals	Calcium, Ca (mg)	36	99	13	14	19	82	9	215	40	138	243	67	109
	Iron, Fe (mg)	0.86	2.71	0.51	0.22	0.43	0.62	0.23	2.32	0.47	6.20	5.08	1.77	1.20
	Magnesium, Mg (mg)	13	79	10	12	17	57	14	55	12	50	80	26	65
	Phosphorus, P (mg)	29	49	28	21	31	61	24	50	26	58	73	48	52
	Potassium, K (mg)	194	558	204	136	296	299	229	611	170	554	569	521	510
	Sodium, Na (mg)	28	79	13	2	5	7	2	20	18	56	31	46	24
	Zinc, Zn (mg)	0.18	0.53	0.07	0.17	0.80	0.58	0.16	0.90	0.18	1.07	1.11	0.50	0.43
Vitamins	Vitamin C, total ascorbic acid (mg)	9.2	28.1	23.4	3.2	84.0	23.0	2.2	43.3	36.6	133.0	31.8	27.0	102.0
	Thiamin (mg)	0.070	0.078	0.060	0.031	0.040	0.200	0.039	0.027	0.061	0.086	0.082	0.067	0.050

Type	Nutrient	Lettuce*	Spinach*	Tomato*	Cucumber*	Bitter gourd*	Okra*	Brinjal/ Eggplant*	Amaranth*	Cabbage*	Parsley*	Peppermint*	Coriander*	Basella*
	Riboflavin (mg)	0.080	0.189	0.040	0.025	0.040	0.060	0.037	0.158	0.040	0.098	0.266	0.162	0.155
	Niacin (mg)	0.375	0.724	0.500	0.037	0.400	1.00	0.647	0.658	0.234	1.313	1.706	1.114	0.500
	Vitamin B-6 (mg)	0.090	0.195	0.081	0.051	0.043	0.215	0.084	0.192	0.124	0.090	0.129	0.146	0.240
	Folate, DFE (µg)	38	194	9.0	14	72	60	22	85	43	152	114	62	140
	Vitamin A, RAE (µg)	370	469	32	4	24	36	1.00	146	5	421	212	337	400
	Vitamin A, IU (IU)	7405	9377	642	72	471	716	23	2917	98	8421	4284	6748	8000
	Vitamin E (alpha-tocopherol) (mg)	0.22	2.03	0.38	0.03	-	0.27	0.30	-	0.15	0.75	-	2.50	-
	Vitamin K (phylloquinone) (µg)	126.3	482.9	10.1	7.2	-	31.3	3.5	1140.0	76.0	1640	-	310.0	-
Lipids	Fatty acids, total saturated (g)	0.020	0.063	0.028	0.013	-	0.026	0.034	0.091	0.034	0.132	0.246	0.014	-
	Fatty acids, total monounsaturated (g)	0.006	0.010	0.030	0.002	-	0.017	0.016	0.076	0.017	0.295	0.033	0.275	-
	Fatty acids, total polyunsaturated (g)	0.082	0.165	0.081	0.003	-	0.027	0.076	0.147	0.017	0.124	0.508	0.040	-

*Per 100 g fresh weight

Adapted from the U.S. Department of Agriculture, Agriculture Research Service (USDA, 2015).

Table 3. Arsenic uptake in the edible parts of fruits and vegetables in relation to soil and irrigation water in some selected arsenic contaminated regions of the world

Country	Plant name	Arsenic concentration			Reference
		Agricultural soil (mg As/kg)	Irrigation water (µg As/L)	Fruits/Vegetables (mg As/kg)	
Fruits					
India	Mango	13.12	410–1010[a]	0.091	(Rahaman et al., 2013)
India	Lychee	13.12	410–1010[a]	0.119	(Rahaman et al., 2013)
India	Papaya	13.12	410–1010[a]	0.123-0.276	(Rahaman et al., 2013)
Bangladesh	Papaya	-	240-1800[a]	0.40	(Alam et al., 2003)
India	Guava	13.12	410–1010[a]	0.02-0.056	(Rahaman et al., 2013)
India	Ber	13.12	410–1010[a]	0.02	(Rahaman et al., 2013)
Vegetables					
Portugal	Lettuce	1.7-1850.8	-	0.2-1.8	(Alvarenga et al., 2014)
Bangladesh	Spinach	0.303-8.628	129-532[a]	0.096-.126	(Farid et al., 2003)
Taiwan	Spinach		13.8-881[a]	0.046	(Kar et al., 2013)
Pakistan	Bitter gourd	6-57	0.015-0.098[b]	0.275-1.11	(Baig and Kazi, 2012)
India	Bitter gourd	13.12	410-1010[a]	0.262	(Rahaman et al., 2013)
Pakistan	Ladies' fingers/Okra		11.3–55.8[a] 8.7–46.2[d]	0.894 0.324	(Arain et al., 2009)
India	Brinjal		110[a]	0.086–0.212	(Roychowdhury et al., 2003)
Pakistan	Brinjal	6-57	0.015-0.098[a]	0.17-0.570	(Baig and Kazi, 2012)
Nepal	Brinjal	-	1014[a]	0.14	(Dahal et al., 2008)
Portugal	Cabbage	1.7-1850.8	-	0.17-0.40	(Alvarenga et al., 2014)

Country	Plant name	Arsenic concentration			Reference
		Agricultural soil (mg As/kg)	Irrigation water (μg As/L)	Fruits/Vegetables (mg As/kg)	
Taiwan	Cabbage	-	13.8-881 [a]	0.0126	(Kar et al., 2013)
India	Cabbage	13.12	410-1010 [a]	0.211–0.456	(Rahaman et al., 2013)
Pakistan	Peppermint	-	11.3–55.8 [c] 8.7–46.2 [d]	1.20 0.45	(Arain et al., 2009)
Pakistan	Coriander	-	11.3–55.8 [c] 8.7–46.2 [d]	0.985 0.185	(Arain et al., 2009)
Taiwan	Basella	-	13.8-881 [a]	0.0181	(Kar et al., 2013)

[a] Groundwater; [b] Tube well water; [c] Lake water; [d] Canal water.

Table 4. Range of arsenic accumulation for various fruits and vegetables estimated from literature data

Fruits and vegetables	Family	Species	Arsenic uptake range (mg/kg)	Reference
Fruits				
Mango	Anacardiaceae	*Mangifera indica* L.	0.09-0.128	(Rahaman et al., 2013; Saha and Zaman, 2013; Liao et al., 2014)
Lychee	Sapindaceae	*Litchi chinensis* Sonn	0.119-0.125	(Rahaman et al., 2013; Saha and Zaman, 2013)
Papaya	Caricaceae	*Carica papaya* L.	0.123-0.4	(Roychowdhury et al., 2002; Alam et al., 2003; Rahaman et al., 2013)
Guava	Myrtaceae	*Psidium guajava* L.	0.028-8.4	(Prieto-García et al., 2005; Rahaman et al., 2013)
Sapota	Sapotaceae	*Manilkara zapota* (L.) P.Royen	2.8	(Prieto-García et al., 2005)
Avocado	Lauraceae	*Persea americana* Mill.	5.7	(Prieto-García et al., 2005)
Ber	Rhamnaceae	*Zizyphus mauritiana* Lam.	0.02	(Rahaman et al., 2013)
Banana	Musaceae	*Musa*spp.	2.54-8.43	(Prieto-García et al., 2005; Ahiamadjie et al., 2011)
Vegetables				
Lettuce	Asteraceae	*Lactuca sativa* L.	0.02-32	(Cao and Ma, 2004; Juhasz et al., 2008; Alvarenga et al., 2014; Caporale et al., 2014)
Spinach	Amaranthaceae	*Spinacia oleracea* L.	0.05-0.9	(Farid et al., 2003; Bhatti et al., 2013; Arain et al., 2014)

Fruits and vegetables	Family	Species	Arsenic uptake range (mg/kg)	Reference
Betel	Piperaceae	*Piper betle* L.	0.094-0.107	(Al-Rmalli et al., 2011; Nema et al., 2014)
Tomato	Solanaceae	*Solanum lycopersicum* L.	0.001-2.36	(Anawar et al., 2012; Mishra et al., 2014)
Cucumber	Cucurbitaceae	*Cucumis sativus* L.	0.020-675.5	(Hong et al., 2011a; Hong et al., 2011b; Anawar et al., 2012)
Pointed gourd	Cucurbitaceae	*Trichosanthes dioica* Roxb.	0.151	(Saha and Zaman, 2013)
Bitter gourd	Cucurbitaceae	*Momordica charantia* L	0.117-1.11	(Baig and Kazi, 2012; Saha and Zaman, 2013)
Okra	Malvaceae	*Abelmoschus esculentus* (L.) Moench	0.073-0.84	(Alam et al., 2003; Samal et al., 2011; Mishra et al., 2014)
Brinjal	Solanaceae	*Solanum melongena* L.	0.19-217	(Roychowdhury et al., 2002; Samal et al., 2011; Anawar et al., 2012)
Amaranth	Amaranthaceae	*Amaranthus acanthociton* I.D.Sauer	0.372-20.81	(Al-Rmalli et al., 2005; Kar et al., 2013; Bian et al., 2014)
Cabbage	Brassicaceae	*Brassica oleracea*	0.17-1.01	(Rahaman et al., 2013; Alvarenga et al., 2014)
Parsley	Apiaceae	*Petroselinum crispum*	0.08-11	(Prieto-García et al., 2005; Atkinson et al., 2012)
Peppermint	Lamiaceae	*Mentha × piperita* L.	0.45-1.20	(Arain et al., 2009; Baig and Kazi, 2012)
Coriander	Apiaceae	*Coriandrum sativum* L.	0.2-2.7	(Prieto-García et al., 2005; Alvarenga et al., 2014)
Basella	Basellaceae	*Basella alba* L.	0.0181	(Kar et al., 2013)

Basella (*Basella alba* L.) is widely consumed as a leafy vegetable and considered as a rich source of vitamin A, vitamin C, iron, and calcium. It has antifungal, anticonvulsant, analgesic, anti-inflammatory, androgenic, antioxidant, and antimutagenic activities (Table 2) (Adhikari et al., 2012). The accumulation of arsenic in Basella was 0.0181 mg/kg a high level of groundwater contamination area of Chianan Plain, Southwestern Taiwan (Table 3) (Kar et al., 2013).

Amaranth (*Amaranthus* sp.) is a cosmopolitan genus of annualor short-lived perennial plants. It is a good source of protein, carbohydrate, vitamins, calcium, and potassium (Table 2) (USDA, 2015). Some amaranth species are cultivated as leaf and stem vegetables, cereals, and ornamental plants. It exhibits lysine, phosphorus, potassium, magnesium. It can protect cardiovascular diseases, cancer and diabetes (Montoya-Rodríguez et al., 2015). The contamination arsenic in amaranth was evaluated in an arsenic-affected area of West Bengal, India and the uptake of arsenic was 0.372 mg/kg (Bhattacharya et al., 2010). The concentrations of arsenic in biogas slurry ranged from 0.30 to 0.5 mg/l. Amaranth was observed to have a greater arsenic absorption capacity (20.81 mg/kg) compared with other vegetables (Bian et al., 2014). The concentration of arsenic varied from 0.04 to 0.33 mg/kg from the area adjacent to the Hazaribag tannery, Dhaka, Bangladesh (Islam et al., 2014). The uptake and translocation of arsenic in the edible part in a groundwater contaminated area in the coastal part of Chianan Plain, Southwestern Taiwan was investigated and the accumulation of arsenic was 0.064 mg/kg (Table 3) (Kar et al., 2013). A market basket survey of arsenic in foodstuffs in the United Kingdom was conducted. The accumulation of arsenic in amaranthus ranged from 0.015 to 0.160 mg/kg (Al-Rmalli et al., 2005). Furthermore, amaranthus was cultivated in an arsenic-affected area in West Bengal India and the accumulation of arsenic was 0.270 mg/kg (Samal et al., 2011).

Tomato (*Solanum lycopersicum* L.), an important source of nourishment, is one of the most important vegetable for the whole world's population. It represents an important source of antioxidants and other bioactive compounds (Table 2) (USDA, 2015). Tomato is an important source of nourishment for the whole world's population. The mean consumption of annual fresh tomato is 18 kg per European and 8 kg per capita in the US (Raiola et al., 2014). Due to its principal carotenoid lycopene content, tomato is an important antioxidant source which may prevent cancer especially prostate cancer (Lin et al., 2015). In general, tomato has the ability to accumulate low level of arsenic. But it also can uptake more than the hygiene level depending on the soil and/or water. The presence of arsenic reduced germination (20-40%) of tomato seed on a nutrient medium (Marmiroli et al., 2014).In fact, increasing arsenic level in irrigation water, the arsenic concentration in the roots was increased. Arsenic in tomato plants was mainly accumulated in roots (85% of total As), followed by shoots (14%) and fruit (1%) when grown in nutrient solution (Burlo et al., 1999). In another pot experiment, it was observed that arsenic in tomato plants was mainly accumulated in roots (78% of total arsenic), followed by shoots (16%) and fruit (6%) in absence of phosphorus, whereas in the presence of phosphorus, arsenic was accumulated less(4%) (Pigna et al., 2012). In addition, the concentration of arsenic in tomato ranged from 0.041to 0.052 mg/kg and 0.001to 0.112 mg/kg collected from Salamanca (Spain) and market of Dhaka, Bangladesh, respectively, was observed by Anawar et al., (2012).. In a recent study, the concentration of arsenic varied from 0.048 to 0.062 mg/kg in the tomato paste and ketchup samples, collected from markets of Tehran, Iran (Hadiani et al., 2014). However, the health risk for adults and adolescents was within acceptable limits with an average weekly intake of 500 g of tomato. A strong linear correlation was found between arsenic in tomato roots (r = 0.95) and soil extractable arsenic, but a weaker correlation was obtained relatively to shoots (r = 0.56) (Madeira et al., 2012). The concentration of arsenic ranged from1.32 to 2.36 mg/kg in tomato cultivated in the Yamuna flood plains of Delhi, India (Mishra et al., 2014). In the coastal part of Chianan Plain, the accumulation of arsenic in tomato was 53.4 µg/kg in groundwater contaminated sites (Kar et al., 2013).

Cucumber (*Cucumis sativas* L.) is a widely cultivated vegetable plant containing glycosphingo lipids. It is a source of vitamin E, calcium and potassium (Table 2) (USDA, 2015).It shows anticancer activities (Sugawara et al., 2006). The uptake of arsenic in fruit vegetable is low comparing with root and leafy vegetable. The uptake and translocation of arsenic on xylem sap focus generally on the concentration and speciation of arsenic in the xylem. Arsenate exposure had a significant influence on sap production, leading to a reduction of up to 96% sap production when plants were exposed to 1000 µg/kgAs(V) (Uroic et al., 2012). It was similar to the previous results (Mihucz et al., 2005). In another study, the effect of arsenite, arsenate and dimethylarsinic acid, on the accumulation of arsenic in cucumber were evaluated. The soils were contaminated with 20 and 50 mg/kg of arsenite, arsenate, or dimethylarsinic acidfor 40 days, the growth was markedly inhibited by the arsenic species. Irrespective of the arsenic species, the arsenic concentrations accumulated in cucumber increased with increasing arsenic concentration in the soil. The range of arsenic uptake was in root 506 to 2096 mg/kg followed by shoot 7.19 to 26.06 mg/kg (Hong et al., 2011a). In addition, phytoextraction is a remediation technology with a promising application for removing arsenic from soils and waters. Several plants were evaluated for their arsenic uptake capacity hydroponically amended with 25 mgAs/l. The germination ratio of cucumber (90.5%) was the highest among the plants tested (60.0, 44.4, and 63.3% for corn, sorghum, and wheat, respectively). Cucumber displayed the highest tolerance against arsenic among the plants. The accumulated arsenic concentrations in the shoot and root were 675.5 and 312.0 mg/kg, respectively suggesting to be a candidate plant for phytoextraction of arsenic from soils and water (Hong et al., 2011b). In a market basket survey from Dhaka, Bangladesh, the uptake of arsenic varied from 0.020 to 0.175 mg/kg in cucumber (Anawar et al., 2012).

Pointed gourd (*Trichosanthes dioica* Roxb.) is another common vegetable containing carbohydrates, vitamin A, vitamin C, magnesium, potassium, copper, sulfur and chlorine mainly consumed in the South Asia. It has antipyretic, diuretic, cardiotonic, laxative and antiulcer activity

(Kumar et al., 2012).Only one record of the accumulation of arsenic in pointed gourd was investigated in a market basket survey. The concentration of arsenic was 0.151 mg/kg in pointed gourd collected from the central market of Rajshahi City, Bangladesh (Saha and Zaman, 2013).

Bitter gourd (*Momordica charantia* L.) is one of the most popular vegetables, generally consumed cooked in the green or early yellowing stage in Southeast Asia. It is a good source of protein, iron, and vitamins (Table 2) (USDA, 2015). It exhibits phytochemicals including triterpenes, proteins, and steroids. It has anti-diabetic, anti-ulcerogenic, anti-mutagenic, antioxidant, anti-lipolytic, analgesic, anti-viral activity (Anilakumar et al., 2015). In a recent study, evaluation of possible health risks of heavy metals including arsenic was investigated. The authors found that the arsenic concentration was 0.117 mg/kg in bitter gourd collected from the central market of Rajshahi City, Bangladesh (Saha and Zaman, 2013). Vegetable was collected randomly from Faiz Ganj, Thari Mirwahand Gambat, sub-district Sindh, Pakistan. Translocation of arsenic contents in vegetables was studied where the soil and the tube well water samples were contaminated by arsenic (Table 3) where the accumulation of arsenic in bitter gourd varied from 0.275 to 1.11 mg/kg (Baig and Kazi, 2012). The uptake of arsenic in different food was investigated. The concentration of arsenic in groundwater was higher than the international permissible limit and the soil arsenic level was little high than the permissible limit (Table 3). The accumulation of arsenic was 0.076 mg/kg in bitter gourd (Rahaman et al., 2013). In a previous study, it was found that bitter gourd accumulated 0.262 mgAs/kg in an arsenic affected area of West Bengal India (Samal et al., 2011). In the area of Sindh Pakistan, the concentration of arsenic in the fruit of bitter gourd was 0.811 and 0.275 mg/kg irrigated with arsenic-contaminated lake and canal water (Arain et al., 2009).

Okra/ladies' finger (*Abelmoschus esculentus* (L.) Moench) is a popular food due to its high fiber, vitamin C, and folate content (Table 2) (USDA, 2015). Okra is also known for being high in antioxidants and it increases immune system and decreased the age-related muscular degeneration, risk of cancer, cardiovascular disease, and cataract formation (Dutta et al.,

2004). Though limited attention has been paid on the arsenic accumulation in this area, some scattered experiments were carried out. For instance, in Gujarat, India okra was grown in mixed industrial effluent irrigated agricultural field, where irrigation water content 5.92 mg As/l. Authors found arsenic concentration was 9.62 mg As/kg in okra.(Tiwari et al., 2011).In another study, the concentration of arsenic was found 0.073 mg/kg in fruits of okra collected from the central market of Rajshahi City, Bangladesh (Saha and Zaman, 2013). Arsenic-contaminated vegetables including okra was evaluated and the arsenic uptake was found 0.019 mg/kg in okra at Samta village in Bangladesh(Alam et al., 2003).The concentration of arsenic by lady's finger was 0.67-0.84 mg/kg in the Yamuna flood plains (YFP) of Delhi, India (Mishra et al., 2014). In west Bengal India, ladies finger was cultivated in an arsenic-contaminated area and the accumulation of As was 182 μg/kg (Samal et al., 2011). In Sindh, Pakistan, vegetables were cultivated with arsenic-contaminated lake and canal water (Table 3) where the concentration of arsenic in the fruit of okra was 0.894 and 0.324 mg/kg, respectively (Arain et al., 2009).

Eggplant/brinjal (*Solanum melongena* L.) is a fruit vegetable containing flavanoids, anthocyanine, ascorbic acid and protein (Table 2) (USDA, 2015).It has antioxidant activity (Kandoliya et al., 2015). Brinjal was sampled in a groundwater contaminated area in Samta village in the Jessore district of Bangladesh. The concentrationof arsenic was 0.19 mg/kg (Table 3) (Alam et al., 2003).In another study, brinjal was collected from the market of Dhaka, Bangladesh where the concentration of arsenic in brinjal was 0.080–0.293 mg/kg (Anawar et al., 2012). In addition, the concentration of arsenic was 0.086 mg/kg in brinjal collected from the central market of Rajshahi City, Bangladesh (Saha and Zaman, 2013). Survey of arsenic in food composites were made from a groundwater arsenic-contaminated area of West Bengal, India (Table 3). Samples were collected from the village people. The contamination of arsenic in brinjal varied from 0.086 to 0.212 mg/kg (Roychowdhury et al., 2003). In Nadia district of West Bengal, India, the accumulation of arsenic in brinjal was 217 mg/kg (Samal et al., 2011). The contamination of arsenic varied from

0.01 to 0.41 mg/kg in brinjal in an arsenic-affected area of West Bengal, India (Bhattacharya et al., 2010). In three subdistricts (Faiz Ganj, Thari Mirwah, Gambat) of Pakistan, translocation of arsenic contents in vegetables was evaluated. The uptake of arsenic in brinjal ranged from 0.17 to 0.570 mg/kg (Baig and Kazi, 2012). In Sindh, Pakistan, Brinjal was cultivated in the contaminated lake and canal water. The concentration of arsenic in fruit was 0.570 and 0.350 mg/kg, respectively (Arain et al., 2009). The influence of arsenic-contaminated irrigation water on the uptake of arsenic in brinjal was investigated at field level in Nepal. The uptake of arsenic was in the brinjal fruit was 0.14 mg/kg (range, <0.10-0.47 mg/kg), irrigated with highly contaminated groundwater (Table 3) (Dahal et al., 2008). High arsenic concentrations (1.03 mg/kg) in brinjal have been reported in superficial water in the Yamuna flood plains (YFP), Delhi, India (Mishra et al., 2014). The translocation of arsenic in brinjal growing under industrial wastewater irrigated an agricultural field of Vadodara, Gujarat, India was evaluated. The uptake of arsenic varied from 0.31 to 0.54 mg/kg (Tiwari et al., 2011).

Coriander (*Coriandrum sativum* L.), also known as Chinese parsleyordhania,is an annual herb. The seeds are often eaten as a spice or an additional ingredient in other foods. It contains linalool, α-pinene, β-pinene, γ-terpinene, geraniol, and camphor. It has antioxidant, anti-inflammatory, anti-diabetic and anticancer activities (Chithra and Leelamma, 2000; Gallagher et al., 2003; Sultana et al., 2010). The accumulation of arsenic by vegetables produced in the vicinity of abandoned pyrite mines, eighteen farms were selected near three mines from the Portuguese sector of the Iberian Pyrite Belt (São Domingos, Aljustrel and Lousal) was evaluated. The uptake of arsenic in coriander (*C. sativum*) varied from 0.20 to 0.30 mg/kg (Alvarenga et al., 2014). In Sindh, Pakistan, coriander was irrigated with arsenic-contaminated lake and canal water. The concentrations of arsenic in leaves were 0.985 and 0.185 irrigated with lake and canal water respectively (Table 3) (Arain et al., 2009). The concentration of arsenic measured in the cultivated plants from Zimapán mining areas, Mexico. The uptake of arsenic was in coriander leaf was 2.7 mg/kg (Prieto-García et al., 2005).

3. HUMAN HEALTH RISK BY ARSENIC CONTAMINATED FRUITS AND VEGETABLES

Arsenic is a non-threshold class one carcinogen. Long term arsenic ingestion, even at low level, could lead to carcinogenesis in humans (Mandal and Suzuki, 2002). The arsenic content of fruits and vegetables is important because people eat large amounts of fruits and vegetables for a healthy diet that is associated with reduced risk for chronic diseases (Wang et al., 2014).

Foods are the major sources of arsenic ingestion in humans who are not exposed to elevated arsenic in drinking water. Arsenic ingestion by fruits and vegetables consumption may not cause a major health problem alone,but when the arsenic-contaminated fruits and vegetable intake by human is added with arsenic-contaminated water for a long time, it is a concern. Because small differences of arsenic in fruits and vegetables can be the additional source of arsenic in the arsenic-contaminated drinking water region. Consequently, it could be a major cause of various carcinogenic and non-carcinogenic diseases.

An investigation was made to detect the levels of arsenic in the composite samples of commonly consumed foodstuffs collected different zones for the first time in Bangladesh. Most of the individual food composites contain a significant level of arsenic where the range was 0.077 to 1.5 mg/kg. The estimated daily dietary intakes clearly exceeded the previous provisional tolerable daily intake value of 2.1 µg/kg-bw/day recommended by the World Health Organization. Fruits and vegetables contribute about 30% to the daily intake of inorganic arsenic (Ahmed et al., 2016). Likewise, estimated daily intake of arsenic by vegetables for Bangladeshi population was approximately 25% (Islam et al., 2014; Joseph et al., 2015) where the lifetime risk of cancer was estimated as 1.15 million people (Joseph et al., 2015).

In Hubei, China, intake of vegetable was the second potential reason for arsenic exposure, responsible for 15.2% of total daily dietary intake. The hazard quotients (HQs) of total daily intake of arsenic was > 1, suggesting direct health effect of arsenic contamination is concern (Cai et al., 2015). Another study was conducted in Xiamen, China to know the concentrations of arsenic and other heavy metals in vegetables, fruits, meat, and seafood samples and their associated health risks were observed. There was an estimated 66% daily intake of arsenic from vegetables. The HQs values for As was < 1, indicating minimum non-carcinogenic risks from arsenic for residents of Xiamen. The projected lifetime excess cancer risk noted that the carcinogenic rate of arsenic is more than the allowable risk level of 10^{-4} (Chen et al., 2011).

The presence of arsenic in vegetable samples collected from agricultural area from Khyber Pakhtunkhwa Province, Pakistan was examined. The mean concentration of arsenic in edible parts of vegetables varied from 0.03 to 1.38 mg/kg. It was observed that the concentrations of arsenic in 75% of the vegetable samples were more than the safe maximum hygienic limit (0.1 mg/kg) set by WHO/FAO. The authors revealed minimal health risk (Hazard index < 1) associated with consumption of vegetables for the local inhabitants. The incremental lifetime cancer risk values for inorganic arsenic indicated a minimal potential cancer risk through ingestion of vegetables. In addition, the hazard quotient values for total arsenic were < 1, indicating minimal non-cancer risk (Rehman et al., 2016).

CONCLUSION

Arsenic in the food chain is a concern. A large number of people particularly from South and Southeast Asia and Latin America are more exposed to arsenic. It is a fundamental right to know the effect of arsenic even at a lower level. Therefore regular monitoring programs for arsenic can inform the update situation. Special attention should be paid to the children due to their sensitivity to arsenic comparing to adults. This review could be helpful for the legislative organization for fixing the level of arsenic for fruits and vegetables.

The findings of arsenic contamination in fruits and vegetables discussed here have an awful sarcasm for the health conscious diet with dairy avoidance and a dependence on fruits and vegetables, which can lead to a greatly increased exposure to a chronic exposure carcinogen (inorganic arsenic). In particular, vegan and macrobiotic diets are of concern (Zhu et al., 2008).

Very few efforts have been undertaken for the remediation of arsenic in fruits and vegetables. As usual, biochar universally reduced concentrations of arsenic in tomato plant organs and tissues compared to the control of contaminated soil. Biochar addition to an arsenic-contaminated soil reduces uptake to tomato plants and in fruits arsenic concentration was very low (<3 µg/kg) (Beesley et al., 2013). Moreover, the inhibitory effect of arsenic was mitigated by the addition of $CaSiO_3$, the provisionary use of $CaSiO_3$was significantly ameliorative in tomato plants (Marmiroli et al., 2014). Research is needed to identify or breed fruits and vegetable cultivars with low accumulation of arsenic in the edible part of the plant.

A good number of studies were carried out for the remediation of arsenic from cereals particularly from rice (Azam et al., 2016b). Those remediation techniques could also be applied for fruits and vegetables.

Field scale studies should be performed to understand the real interaction of arsenic and phosphorus. Studies on arsenic bioaccesibility are needed for determining human As intake from fruits and vegetables for use in the accurate risk assessments to establish updated legislation regarding the maximum level of arsenic in food.

ACKNOWLEDGMENTS

The authors wish to thank Mr. Sanowar Hossain Faruque for extensively editing the manuscript.

REFERENCES

Adhikari, R., Naveen Kumar, H.N., Shruthi, S.D., 2012. A review on medicinal importance of *Basella alba* L. *Int. J. Pharma. Sci. Drug Res* 4, 110-114.

Ahiamadjie, H., Serfor-Armah, Y., Tandoh, J.B., Gyampo, O., Ofosu, F.G., Dampare, S.B., Adotey, D.K., Nyarko, B.J., 2011. Evaluation of trace elements contents in staple foodstuffs from the gold mining areas in southwestern part of Ghana using neutron activation analysis. *J. Radioanal. Nucl. Chem.* 288, 653-661.

Ahmed, M.K., Shaheen, N., Islam, M.S., Habibullah-Al-Mamun, M., Islam, S., Islam, M.M., Kundu, G.K., Bhattacharjee, L., 2016. A comprehensive assessment of arsenic in commonly consumed foodstuffs to evaluate the potential health risk in Bangladesh. *Sci. Total. Environ.* 544, 125-133.

Al-Rmalli, S.W., Haris, P.I., Harrington, C.F., Ayub, M., 2005. A survey of arsenic in foodstuffs on sale in the United Kingdom and imported from Bangladesh. *Sci. Total. Environ.* 337, 23-30.

Al-Rmalli, S.W., Jenkins, R.O., Haris, P.I., 2011. Betel quid chewing elevates human exposure to arsenic, cadmium and lead. *J. Hazard. Mater.* 190, 69-74.

Alam, M.G.M., Snow, E.T., Tanaka, A., 2003. Arsenic and heavy metal contamination of vegetables grown in Samta village, Bangladesh. *Sci. Total. Environ.* 308, 83-96.

Alvarenga, P., Simões, I., Palma, P., Amaral, O., Matos, J.X., 2014. Field study on the accumulation of trace elements by vegetables produced in the vicinity of abandoned pyrite mines. *Sci. Total. Environ.* 470, 1233-1242.

Anawar, H.M., Garcia-Sanchez, A., Hossain, M.N., Akter, S., 2012. Evaluation of health risk and arsenic levels in vegetables sold in markets of Dhaka (Bangladesh) and Salamanca (Spain) by hydride generation atomic absorption spectroscopy. *Bull. Environ. Contam. Toxicol.* 89, 620-625.

Anilakumar, K.R., Kumar, G.P., Ilaiyaraja, N., 2015. Nutritional, Pharmacological and Medicinal Properties of Momordica Charantia. *Int. J. Food. Sci. Nutr.* 4, 75.

Arain, M.B., Kazi, T.G., Baig, J.A., Jamali, M.K., Afridi, H.I., Shah, A.Q., Jalbani, N., Sarfraz, R.A., 2009. Determination of arsenic levels in lake water, sediment, and foodstuff from selected area of Sindh, Pakistan: estimation of daily dietary intake. *Food. Chem. Toxicol.* 47, 242-248.

Arain, S.S., Kazi, T.G., Afridi, H.I., Brahman, K.D., Shah, F., Mughal, M.A., 2014. Arsenic content in smokeless tobacco products consumed by the population of Pakistan: related health risk. *J. AOAC Int.* 97, 1662-1669.

Aruoma, O.I., Sun, B., Fujii, H., Neergheen, V.S., Bahorun, T., Kang, K.S., Sung, M.K., 2006. Low molecular proanthocyanidin dietary biofactor Oligonol: Its modulation of oxidative stress, bioefficacy, neuroprotection, food application and chemoprevention potentials. *Biofactors* 27, 245-265.

Atkinson, N.R., Young, S.D., Tye, A.M., Breward, N., Bailey, E.H., 2012. Does returning sites of historic peri-urban waste disposal to vegetable production pose a risk to human health?-A case study near Manchester, UK. *Soil Use Manage.* 28, 559-570.

Azam, S.M.G.G., Afrin, S., Naz, S., 2016a. Arsenic in cereals, their relation with human health risk and possible mitigation strategies, *Food Rev. Int.* 33, 620-643.

Azam, S.M.G.G., Sarker, T.C., Naz, S., 2016b. Factors affecting the soil arsenic bioavailability, accumulation in rice and risk to human health: a review. *Toxicol Mech Methods* 26, 565-579.

Baig, J.A., Kazi, T.G., 2012. Translocation of arsenic contents in vegetables from growing media of contaminated areas. *Ecotoxicol. Environ. Saf.* 75, 27-32.

Bazzano, L.A., 2005. Dietary intake of fruit and vegetables and risk of diabetes mellitus and cardiovascular diseases [electronic resource], in *Background Paper of the Joint FAO/WHO Workshop on Fruit and Vegetables for Health*, World Health Organization, Kobe, Japan, pp. 1-65.

Beesley, L., Marmiroli, M., Pagano, L., Pigoni, V., Fellet, G., Fresno, T., Vamerali, T., Bandiera, M., Marmiroli, N., 2013. Biochar addition to an arsenic contaminated soil increases arsenic concentrations in the pore water but reduces uptake to tomato plants (*Solanum lycopersicum* L.). *Sci. Total. Environ.* 454, 598-603.

Bergqvist, C., Herbert, R., Persson, I., Greger, M., 2014. Plants influence on arsenic availability and speciation in the rhizosphere, roots and shoots of three different vegetables. *Environ. Pollut.* 184, 540-546.

Bhattacharya, P., Samal, A.C., Majumdar, J., Santra, S.C., 2010. Arsenic contamination in rice, wheat, pulses, and vegetables: a study in an arsenic affected area of West Bengal, India. *Water Air Soil Pollut.* 213, 3-13.

Bhatti, S.M., Anderson, C.W.N., Stewart, R.B., Robinson, B.H., 2013. Risk assessment of vegetables irrigated with arsenic-contaminated water. *Env. Sci. Proces.s Impact* 15, 1866-1875.

Bian, B., Lv, L., Yang, D., Zhou, L., 2014. Migration of heavy metals in vegetable farmlands amended with biogas slurry in the Taihu Basin, China. *Ecol. Eng.* 71, 380-383.

Biswas, A., Biswas, S., Santra, S.C., 2012. Risk from winter vegetables and pulses produced in arsenic endemic areas of Nadia district: field study comparison with market basket survey. *Bull. Environ. Contam. Toxicol.* 88, 909-914.

Bohari, Y., Lobos, G., Pinochet, H., Pannier, F., Astruc, A., Potin-Gautier, M., 2002. Speciation of arsenic in plants by HPLC-HG-AFS: extraction optimisation on CRM materials and application to cultivated samples. *J. Environ. Monit.* 4, 596-602.

Bundschuh, J., Nath, B., Bhattacharya, P., Liu, C.W., Armienta, M.A., Lopez, M.V.M., Lopez, D.L., Jean, J.S., Cornejo, L., Macedo, L.F.L., 2012. Arsenic in the human food chain: the Latin American perspective. *Sci. Total. Environ.* 429, 92-106.

Burlo, F., Guijarro, I., Carbonell-Barrachina, A.A., Valero, D., Martinez-Sanchez, F., 1999. Arsenic species: effects on and accumulation by tomato plants. *J. Agri. Food Chem.* 47, 1247-1253.

Cai, L.M., Xu, Z.C., Qi, J.Y., Feng, Z.Z., Xiang, T.S., 2015. Assessment of exposure to heavy metals and health risks among residents near Tonglushan mine in Hubei, China. *Chemosphere* 127, 127-135.

Cao, X., Ma, L.Q., 2004. Effects of compost and phosphate on plant arsenic accumulation from soils near pressure-treated wood. *Environ. Pollut.* 132, 435-442.

Caporale, A.G., Sommella, A., Lorito, M., Lombardi, N., Azam, S.M.G.G., Pigna, M., Ruocco, M., 2014. Trichoderma spp. alleviate phytotoxicity in lettuce plants (*Lactuca sativa* L.) irrigated with arsenic-contaminated water. *J. Plant. Physiol.* 171, 1378-1384.

Chang, C.Y., Yu, H.Y., Chen, J.J., Li, F.B., Zhang, H.H., Liu, C.P., 2014. Accumulation of heavy metals in leaf vegetables from agricultural soils and associated potential health risks in the Pearl River Delta, South China. *Environ. Monit. Assess.* 186, 1547-1560.

Chen, C., Qian, Y., Chen, Q., Li, C., 2011. Assessment of daily intake of toxic elements due to consumption of vegetables, fruits, meat, and seafood by inhabitants of Xiamen, China. *J. Food Sci.* 76, T181-T188.

Chithra, V., Leelamma, S., 2000. Coriandrum sativum — effect on lipid metabolism in 1, 2-dimethyl hydrazine induced colon cancer. *J. Ethnopharmacol.* 71, 457-463.

Cobb, G.P., Sands, K., Waters, M., Wixson, B.G., Dorward-King, E., 2000. Accumulation of heavy metals by vegetables grown in mine wastes. *Environ. Toxicol. Chem.* 19, 600-607.

Dahal, B.M., Fuerhacker, M., Mentler, A., Karki, K.B., Shrestha, R.R., Blum, W.E.H., 2008. Arsenic contamination of soils and agricultural plants through irrigation water in Nepal. *Environ. Pollut.* 155, 157-163.

DuPont, M.S., Mondin, Z., Williamson, G., Price, K.R., 2000. Effect of variety, processing, and storage on the flavonoid glycoside content and composition of lettuce and endive. *J. Agri. Food Chem.* 48, 3957-3964.

Dutta, D., Chaudhuri, U.R., Chakraborty, R., 2004. Structure, health benefits, antioxidant property and processing and storage of carotenoids. *Afr. J. Food Agric. Nutr. Dev.* 4, 1510-1520.

Ebeid, H.M., Gibriel, A.A.Y., Al-Sayed, H.M.A., Elbehairy, S.A., Motawe, E.H., 2015. Hepatoprotective and Antioxidant Effects of Wheat, Carrot, and Mango as Nutraceutical Agents against CCl4-Induced Hepatocellular Toxicity. *J. Am. Coll. Nutr.*, 1-4.

Farid, A.T.M., Roy, K.C., Hossain, K.M., Sen, R., 2003. Study of arsenic contaminated irrigationwater and it's carried over effect on vegetable. In: Ahmed, M.F., Ali, M.A., Adeel, Z. (Eds.), *Fate of Arsenic in the Environment. BUET/UNU*, Dhaka, Bangladesh, pp. 113–121.

Gallagher, A.M., Flatt, P.R., Duffy, G., Abdel-Wahab, Y.H.A., 2003. The effects of traditional antidiabetic plants on in vitro glucose diffusion. *Nutr. Res.* 23, 413-424.

Gironés-Vilaplana, A., Valentão, P., Andrade, P.B., Ferreres, F., Moreno, D.A., García-Viguer, C., 2015. Beverages of lemon juice and exotic noni and papaya with potential for anticholinergic effects. *Food Chem.* 170, 16-21.

Gutiérrez, R.M.P., Mitchell, S., Solis, R.V., 2008. Psidium guajava: a review of its traditional uses, phytochemistry and pharmacology. *J. Ethnopharmacol.* 117, 1-27.

Hadiani, M.R., Farhangi, R., Soleimani, H., Rastegar, H., Cheraghali, A.M., 2014. Evaluation of heavy metals contamination in Iranian foodstuffs: canned tomato paste and tomato sauce (ketchup). *Food Addit. Contam. Part B Surveill.* 7, 74-78.

Hoffmann, K., Boeing, H., Volatier, J.L., Becker, W., 2003. Evaluating the potential health gain of the World Health Organization's recommendation concerning vegetable and fruit consumption. *Public Health Nutr.* 6, 765-772.

Hong, S.H., Choi, S.A., Lee, M.H., Min, B.R., Yoon, C., Yoon, H., Cho, K.S., 2011a. Effect of arsenic species on the growth and arsenic accumulation in *Cucumis sativus. Environ. Geochem. Health* 33, 41-47.

Hong, S.H., Choi, S.A., Yoon, H., Cho, K.S., 2011b. Screening of Cucumis sativus as a new arsenic-accumulating plant and its arsenic accumulation in hydroponic culture. *Environ. Geochem. Health* 33, 143-149.

Hossain, M.F., 2006. Arsenic contamination in Bangladesh - an overview. *Agric. Ecosyst. Environ.* 113, 1-16.

Huang, R.Q., Gao, S.F., Wang, W.L., Staunton, S., Wang, G., 2006. Soil arsenic availability and the transfer of soil arsenic to crops in suburban areas in Fujian Province, southeast China. *Sci. Total Environ.* 368, 531-541.

III, V.L.F., Dreher, M., Davenport, A.J., 2013. *Avocado consumption is associated with better diet quality and nutrient intake, and lower metabolic syndrome risk in US adults: results from the National Health and Nutrition Examination Survey* (NHANES) 2001-2008.

Islam, G.M.R., Khan, F.E., Hoque, M.M., Jolly, Y.N., 2014. Consumption of unsafe food in the adjacent area of Hazaribag tannery campus and Buriganga River embankments of Bangladesh: heavy metal contamination. *Environ. Monit. Assess.* 186, 7233-7244.

Jiang, Y., Zeng, X., Fan, X., Chao, S., Zhu, M., Cao, H., 2015. Levels of arsenic pollution in daily foodstuffs and soils and its associated human health risk in a town in Jiangsu Province, China. *Ecotoxicol. Environ. Saf.* 122, 198-204.

Joseph, T., Dubey, B., McBean, E.A., 2015. Human health risk assessment from arsenic exposures in Bangladesh. *Sci. Total Environ.* 527-528, 552–560.

Juhasz, A.L., Smith, E., Weber, J., Rees, M., Rofe, A., Kuchel, T., Sansom, L., Naidu, R., 2008. Application of an in vivo swine model for the determination of arsenic bioavailability in contaminated vegetables. *Chemosphere* 71, 1963-1969.

Kabir, F., Tow, W.W., Hamauzu, Y., Katayama, S., Tanaka, S., Nakamura, S., 2015. Antioxidant and cytoprotective activities of extracts prepared from fruit and vegetable wastes and by-products. *Food Chem.* 167, 358-362.

Kamboh, A.A., Arain, M.A., Mughal, M.J., Zaman, A., Arain, Z.M., Soomro, A.H., 2015. Flavonoids: health promoting phytochemicals for animal production a review. *J. Anim. Health Prod.* 3, 6-13.

Kandoliya, U.K., Bajaniya, V.K., Bhadja, N.K., Bodar, N.P., Golakiya, B.A., 2015. Antioxidant and Nutritional Components of Egg plant

(Solanum melongena L) Fruit Grown in Saurastra Region. *Int J Curr Microbiol App Sci* 4, 806-813.

Kar, S., Das, S., Jean, J.S., Chakraborty, S., Liu, C.C., 2013. Arsenic in the water-soil-plant system and the potential health risks in the coastal part of Chianan Plain, Southwestern Taiwan. *J. Asian Earth Sci.* 77, 295-302.

Karadaş, C., Kara, D., 2012. Chemometric approach to evaluate trace metal concentrations in some spices and herbs. *Food Chem.* 130, 196-202.

Khanna, R., MacDonald, J.K., Levesque, B.G., 2014. Peppermint oil for the treatment of irritable bowel syndrome: a systematic review and meta-analysis. *J. Clin. Gastroenterol.* 48, 505-512.

Kumar, N., Singh, S., Manvi, R.G., 2012. Trichosanthes dioica Roxb: An overview. *Pharmacogn. Rev.* 6, 61.

Li, Y., Wang, H., Wang, H., Yin, F., Yang, X., Hu, Y., 2014. Heavy metal pollution in vegetables grown in the vicinity of a multi-metal mining area in Gejiu, China: total concentrations, speciation analysis, and health risk. *Environ. Sci. Pollut. Res.* 21, 12569-12582.

Liao, X., Fu, Y., He, Y., Yang, Y., 2014. Occurrence of arsenic in fruit of mango plant (*Mangifera indica* L.) and its relationship to soil properties. *Catena* 113, 213-218.

Lin, P.H., Aronson, W., Freedland, S.J., 2015. Nutrition, dietary interventions and prostate cancer: the latest evidence. *BMC medicine* 13, 3.

Llorach, R., Martínez-Sánchez, A., Tomás-Barberán, F.A., Gil, M.I., Ferreres, F., 2008. Characterisation of polyphenols and antioxidant properties of five lettuce varieties and escarole. *Food Chem.* 108, 1028-1038.

Lomnitski, L., Bergman, M., Nyska, A., Ben-Shaul, V., Grossman, S., 2003. Composition, efficacy, and safety of spinach extracts. *Nutr Cancer* 46, 222-231.

MacLean, K.S., Langille, W.M., 1981. Arsenic in orchard and potato soils and plant tissue. *Plant Soil* 61, 413-418.

Madeira, A.C., De Varennes, A., Abreu, M.M., Esteves, C., Magalhães, M.C.F., 2012. Tomato and parsley growth, arsenic uptake and translocation in a contaminated amended soil. *J. Geochem. Explor.* 123, 114-121.

Mandal, B.K., Suzuki, K.T., 2002. Arsenic round the world: a review. *Talanta* 58, 201-235.

Marmiroli, M., Pigoni, V., Savo-Sardaro, M.L., Marmiroli, N., 2014. The effect of silicon on the uptake and translocation of arsenic in tomato (*Solanum lycopersicum* L.). *Environ. Exper. Bot.* 99, 9-17.

Matschullat, J.r., 2000. Arsenic in the geosphere-a review. *Sci. Total. Environ.* 249, 297-312.

McBride, M.B., 2013. Arsenic and lead uptake by vegetable crops grown on historically contaminated orchard soils. *Appl. Environ. Soil Sci.* 2013.

Mihucz, V.G., Tatár, E., Virág, I., Cseh, E., Fodor, F., Záray, G., 2005. Arsenic speciation in xylem sap of cucumber (*Cucumis sativus* L.). *Anal Bioanal. Chem.* 383, 461-466.

Mishra, B.K., Dubey, C.S., Shukla, D.P., Bhattacharya, P., Usham, A.L., 2014. Concentration of arsenic by selected vegetables cultivated in the Yamuna flood plains (YFP) of Delhi, India. *Environ. Earth Sci.* 72, 3281-3291.

Montoya-Rodríguez, A., Gómez-Favela, M.A., Reyes-Moreno, C., Milán-Carrillo, J., González de Mejía, E., 2015. Identification of Bioactive Peptide Sequences from Amaranth (Amaranthus hypochondriacus) Seed Proteins and Their Potential Role in the Prevention of Chronic Diseases. *Compr. Rev. Food Sci. Food Saf.* 14, 139-158.

Moyano, A., Garcia-Sanchez, A., Mayorga, P., Anawar, H.M., Alvarez-Ayuso, E., 2009. Impact of irrigation with arsenic-rich groundwater on soils and crops. *J Environ Monit* 11, 498-502.

Murcott, S., 2012. *Arsenic contamination in the world: an international sourcebook.* IWA Publishing.

Nakwanit, S., Visoottiviseth, P., Khokiattiwong, S., Sangchoom, W., 2011. Management of arsenic-accumulated waste from constructed wetland treatment of mountain tap-water. *J. Hazard Mater.* 185, 1081-1085.

Nema, N.K., Maity, N., Sarkar, B.K., Mukherjee, P.K., 2014. Determination of trace and heavy metals in some commonly used medicinal herbs in Ayurveda. *Toxicol. Ind. Health* 30, 964-968.

Norton, S.A., 1998. Betel: consumption and consequences. *J. Am. Acad. Dermatol.* 38, 81-88.

Nriagu, J.O., Bhattacharya, P., Mukherjee, A.B., Bundschuh, J., Zevenhoven, R., Loeppert, R.H., 2007. Arsenic in soil and groundwater: an overview. *Trace Metals Other Contamin. Environ.* 9, 3-60.

Paranjpe, R., Gundala, S.R., Lakshminarayana, N., Sagwal, A., Asif, G., Pandey, A., Aneja, R., 2013. Piper betel leaf extract: anticancer benefits and bio-guided fractionation to identify active principles for prostate cancer management. *Carcinogenesis*, bgt066.

Pereira, A., Maraschin, M., 2015. Banana (*Musa* spp) from peel to pulp: Ethnopharmacology, source of bioactive compounds and its relevance for human health. *J. Ethnopharmacol.* 160, 149-163.

Pigna, M., Caporale, A.G., Cozzolino, V., Fernández López, C., Mora, M.L., Sommella, A., Violante, A., 2012. Influence of phosphorus on the arsenic uptake by tomato (*Solanum lycopersicum* L) irrigated with arsenic solutions at four different concentrations. *J. Soil Sci. Plant Nutr.* 12, 775-784.

Porrini, M., Riso, P., Oriani, G., 2002. Spinach and tomato consumption increases lymphocyte DNA resistance to oxidative stress but this is not related to cell carotenoid concentrations. *Eur. J. Nutr.* 41, 95-100.

Prieto-García, F., Callejas, H., Lechuga Mde los, Á., Gaytán, J., EE., B., 2005. Accumulation in vegetable weavings of arsenic originating from water and floors of Zimapán, Hidalgo State, Mexico. *Bioagro* 17, 129-136.

Qi, S., Huang, H., Huang, J., Wang, Q., Wei, Q., 2015. Lychee (*Litchi chinensis* Sonn.) seed water extract as potential antioxidant and anti-obese natural additive in meat products. *Food Control.* 50, 195-201.

Rahaman, S., Sinha, A.C., Pati, R., Mukhopadhyay, D., 2013. Arsenic contamination: a potential hazard to the affected areas of West Bengal, India. *Environ. Geochem. Health* 35, 119-132.

Rahman, M.A., Hasegawa, H., 2011. High levels of inorganic arsenic in rice in areas where arsenic-contaminated water is used for irrigation and cooking. *Sci. Total Environ.* 409, 4645-4655.

Rahman, M.M., Asaduzzaman, M., Naidu, R., 2013. Consumption of arsenic and other elements from vegetables and drinking water from an arsenic-contaminated area of Bangladesh. *J. Hazard. Mater.* 262, 1056-1063.

Raiola, A., Rigano, M.M., Calafiore, R., Frusciante, L., Barone, A., 2014. Enhancing the Health-Promoting Effects of Tomato Fruit for Biofortified Food. *Mediators Inflamm.* 2014.

Rehman, Z.U., Khan, S., Qin, K., Brusseau, M.L., Shah, M.T., Din, I., 2016. Quantification of inorganic arsenic exposure and cancer risk via consumption of vegetables in southern selected districts of Pakistan. *Sci. Total Environ.* 550, 321-329.

Riachi, L.G., De Maria, C.A.B., 2015. Peppermint antioxidants revisited. *Food Chem.* 176, 72-81.

Rodriguez-Amaya, D.B., 2015. Status of carotenoid analytical methods and in vitro assays for the assessment of food quality and health effects. *Curr. Opin. Food Sci.* 1, 56-63.

Roychowdhury, T., Tokunaga, H., Ando, M., 2003. Survey of arsenic and other heavy metals in food composites and drinking water and estimation of dietary intake by the villagers from an arsenic-affected area of West Bengal, India. *Sci. Total Environ.* 308, 15-35.

Roychowdhury, T., Uchino, T., Tokunaga, H., Ando, M., 2002. Survey of arsenic in food composites from an arsenic-affected area of West Bengal, India. *Food Chem. Toxicol.* 40, 1611-1621.

Saha, N., Zaman, M.R., 2013. Evaluation of possible health risks of heavy metals by consumption of foodstuffs available in the central market of Rajshahi City, Bangladesh. *Environ. Monit. Assess.* 185, 3867-3878.

Samal, A.C., Kar, S., Bhattacharya, P., Santra, S.C., 2011. Human exposure to arsenic through foodstuffs cultivated using arsenic contaminated groundwater in areas of West Bengal, India. *J. Environ. Sci. Health A Tox. Hazard Subst. Environ. Eng.* 46, 1259-1265.

Shaibur, M.R., Kawai, S., 2010. Effect of arsenic on nutritional composition of Japanese mustard Spinach: An Ill effect of arsenic on nutritional quality of a green leafy vegetable. *Nat. Sci.* 8, 186-194.

Smith, E., Juhasz, A.L., Weber, J., 2009. Arsenic uptake and speciation in vegetables grown under greenhouse conditions. *Environ. Geochem. Health* 31, 125-132.

Sugawara, T., Zaima, N., Yamamoto, A., Sakai, S., Noguchi, R., Hirata, T., 2006. Isolation of sphingoid bases of sea cucumber cerebrosides and their cytotoxicity against human colon cancer cells. *Biosci. Biotechnol. Biochem.* 70, 2906-2912.

Sultana, S., Ripa, F.A., Hamid, K., 2010. Comparative antioxidant activity study of some commonly used spices in Bangladesh. *Pakistan J. Biol. Sci.* 13, 340.

Tiwari, K.K., Singh, N.K., Patel, M.P., Tiwari, M.R., Rai, U.N., 2011. Metal contamination of soil and translocation in vegetables growing under industrial wastewater irrigated agricultural field of Vadodara, Gujarat, India. *Ecotoxicol. Environ. Saf.* 74, 1670-1677.

Uroic, M.K., Salaün, P., Raab, A., Feldmann, J., 2012. Arsenate impact on the metabolite profile, production, and arsenic loading of xylem sap in cucumbers (*Cucumis sativus* L.). *Front Physiol* 3, 1-23.

USDA, 2015. Agricultural Research Service, Nutrient Data Laboratory. USDA National Nutrient Database for Standard Reference, Release 28. Version Current: September 2015. Internet: http://www.ars.usda.gov/nea/bhnrc/ndl.

Wang, L., Bordi, P.L., Fleming, J.A., Hill, A.M., Kris-Etherton, P.M., 2015. Effect of a moderate fat diet with and without avocados on lipoprotein particle number, size and subclasses in overweight and obese adults: a randomized, controlled trial. *J. Am. Heart Assoc.* 4, e001355.

Wang, X., Ouyang, Y., Liu, J., Zhu, M., Zhao, G., Bao, W., Hu, F.B., 2014. Fruit and vegetable consumption and mortality from all causes, cardiovascular disease, and cancer: systematic review and dose-response meta-analysis of prospective cohort studies. *BMJ* 349, g4490.

Wang, Y., Qiao, M., Liu, Y., Zhu, Y., 2012. Health risk assessment of heavy metals in soils and vegetables from wastewater irrigated area, Beijing-Tianjin city cluster, China. *J. Environ. Sci.* 24, 690-698.

Warren, G.P., Alloway, B.J., Lepp, N.W., Singh, B., Bochereau, F.J.M., Penny, C., 2003. Field trials to assess the uptake of arsenic by vegetables from contaminated soils and soil remediation with iron oxides. *Sci. Total. Environ.* 311, 19-33.

Williams, P.N., Price, A.H., Raab, A., Hossain, S.A., Feldmann, J., Meharg, A.A., 2005. Variation in arsenic speciation and concentration in paddy rice related to dietary exposure. *Environ. Sci. Technol.* 39, 5531-5540.

Zhu, Y.G., Williams, P.N., Meharg, A.A., 2008. Exposure to inorganic arsenic from rice: a global health issue? *Environ. Pollut.* 154, 169-171.

In: Arsenic
Editor: Ratko Knežević

ISBN: 978-1-53612-461-3
© 2017 Nova Science Publishers, Inc.

Chapter 3

DEVELOPMENT OF A WATER TREATMENT PLANT FOR ARSENIC REMOVAL BASED ON THE ZERO-VALENT IRON TECHNOLOGY

Eliana Berardozzi[1,2] and Fernando S. García Einschlag[1,]*
[1]Instituto de Investigaciones Fisicoquímica Teóricas y Aplicadas (INIFTA), CCT-La Plata-CONICET, Departamento de Química, Facultad de Ciencias Exactas, UNLP, Buenos Aires, Argentina
[2]Departamento de Hidráulica, Facultad de Ingeniería, UNLP, Buenos Aires, Argentina

ABSTRACT

Arsenic (As) strongly limits water potability due to its high toxicity. The Chaco-Pampean plain is one of the regions worldwide recognized for its high arsenic content in groundwater and the involved area covers about 106 km^2 in Argentina. Arsenic levels in groundwater above 100 µg/L have been frequently reported. For this reason, it is imperative to develop efficient and inexpensive technical solutions for the elimination of arsenic from drinking water.

[*]Corresponding author: fgarciae@química.unlp.edu.ar.

In the present work a three-module continuous plant design, capable of deliver up to 1 m³/day of arsenic-free drinking water, is described. The system, whose first and main stage is a Zero-Valent Iron (ZVI) reactive bed, is simple, easy to use, was designed to respond small communities' needs and can be adapted for groundwater with different physicochemical characteristics.

Arsenic removal by ZVI-based technologies is related to the corrosion products generated by metallic iron oxidation and involves different mechanisms, including adsorption, surface complexation, surface precipitation and co-precipitation. Iron corrosion rates depend on both the operating conditions and the ZVI source used. Therefore, the effects of changing the main operational variables, of columns packed with iron wool, were analyzed in order to select the most favorable settings.

In many cases, the application of ZVI-based techniques is limited by reactivity losses and reductions of hydraulic conductivity caused by the accumulation of corrosion products. These problems arise due to the formation a thick layer of iron oxides onto ZVI surface, especially in natural waters with relatively high dissolved oxygen content. Consequently, the hydraulic behavior of the designed plant was studied throughout the operation period. In addition, stimulus response tests were carried out periodically to determine the residence time distribution along the reactive column.

The results obtained show that the lifespan of the plant may be predicted by taking in to account both the main chemical processes involved and the fluid dynamic properties of the bed.

Keywords: ZVI-based plant for As removal, operating conditions, hydrodynamic behavior

INTRODUCTION

Iron-Based Technologies for Environmental Remediation

The high deterioration of natural resources in recent decades is of global concern. Legislations, aimed at reinforcing the clean-up, protection and remediation of contaminated land and water sources, have consequently been created. Given that several traditional treatment methods are not sustainable in many situations, the research into alternative technologies for water treatment has increased. A number of these alternative techniques utilize iron and its mineral products to remove or stabilize different kinds of contaminants [1].

Iron is one of the most abundant elements in the earth's crust and, in nature, it predominates in two valence states: Fe(II) and Fe(III). Fe(0) also may be found under specific environmental and geological conditions. Reactions related to this metal play an important role in a number of environmental processes. In this context, different mineral forms of iron influence the mobility, distribution and degradation of pollutants due to its ability to act as reducing, adsorbing or precipitating agents. Amorphous and freshly formed iron oxyhydroxides are known to be particularly effective adsorbents of a wide range of contaminants [2–4].

Applications of iron-based technologies in contaminated or environmental remediation can be divided into two groups: i) technologies which use iron as a sorbent, (co-)precipitant or contaminant immobilizing agent, and ii) those which use iron as an electron donor to chemically reduce the contaminant, however many technologies utilize both processes.

Adsorption is associated with the accumulation of pollutant onto a solid surface, whereas precipitation is due to the spontaneous formation of precipitates. Co-precipitation is an unspecific removal mechanism in which foreign species are entrapped within the matrix of precipitating corrosion products[5–7].Chemical reduction refers to a reaction in which the contaminant gains one or more electrons and is transformed into less toxic or mobile forms[5].

Arsenic Occurrence in Natural Waters

High arsenic content in natural waters is a worldwide problem. Arsenic pollution has been reported in China, Japan, India, Bangladesh, Taiwan, New Zeeland, Poland, Hungary, Canada, USA, Mexico, Chile and Argentina [8]. Long-term drinking water exposure with this contaminant causes skin, lung, bladder, and kidney cancer as well as pigmentation changes, skin thickening, neurological disorders, muscular weakness, loss of appetite, and nausea.The WHO guideline of 10 ppb (0.01 mg/L) has been adopted as the maximum limit allowable for drinking water [9]. However, many countries have retained the previous WHO limit of 50 ppb (0.05 mg/L) as their standard [10].

The Chaco–Pampean plain, in Latin-America, is among the largest identified areas in the world with high arsenic contents in groundwater [11, 12]. Argentina was the first country in Latin America from where the occurrence of arsenic in groundwater has been reported. The area covers about 10^6 km^2 of this country and it is currently estimated that the population living in areas with As-contaminated groundwater rises to about 4 million people [10–12].

Most of the As content in groundwater of the Chaco-Pampean plain has a natural origin. Arsenic is present in most rocks and is mobilized by natural weathering reactions, biological activity, geochemical reactions and volcanic emissions. Despite environmental arsenic problems mainly result from its mobilization under natural conditions, some anthropogenic activities such as mining, combustion of fossil fuels, use of arsenic pesticides, herbicides, and crop desiccants could create additional impacts [13].

Arsenic exists in the −3, 0, +3 and +5 oxidation states [12]. Natural occurring forms include arsenious and arsenic acids, arsenites, arsenates, methylarsenic and dimethylarsenic acids and arsine. Inorganic arsenic species are often found in water supplies, the two forms most common for natural waters being arsenite (AsO_3^{-3}) and arsenate (AsO_4^{-3}). Pentavalent species predominate and are stable in oxygen rich aerobic environments, whereas the trivalent arsenites predominate in moderately reducing anaerobic environments. The prevailing redox conditions for the Argentinean aquifers range from moderately reducing to oxidizing, with As(V) being the major arsenic species in the contaminated groundwater.

The fulfillment of the allowable limit of arsenic in drinking water has been a gradual process in developing countries. Frequently, the allowed limits have been fixed according to the affordable technology in each country. The problem of arsenic-contaminated drinking water has been mitigated, in some mid-sized and large urban areas of Latin America with centralized water supply, by the installation of treatment plants or tapping alternative water resources. However, few solutions are available for arsenic mitigation in rural and small populations. Small-scale and household methods to remove arsenic from water for drinking purposes in Latin America have been recently reviewed [10]. For this population sectors, the development of alternative and inexpensive technologies is of foremost importance.

Mechanisms of Arsenic Removal in Zero-Valent Iron Systems

Among the alternative techniques introduced in the last years are those based on the use of zero-valent iron (ZVI). These technologies have appeared as promising solution tools because they may be implemented by using simple and widely available materials. Their use was initially related to applications in subsurface permeable reactive barriers (ZVI PRBs) for in-situ groundwater remediation [14]. These have been installed worldwide and, in most cases, worked satisfactorily [15, 16]. Hence, the technology was extended for use in ex-situ filter systems.

The mechanisms of As removal by this technique involve sequential stages and Fe species in different oxidation states. In the first step, ZVI is oxidized to Fe(II) by different mechanisms depending whether the treatment was designed to be used under aerobic or anaerobic conditions. For oxygen rich waters, ZVI oxidation is coupled to O_2 reduction which yields OH^-, whereas in anoxic media ZVI oxidation produces H_2 through reduction of H^+. The latter reaction becomes increasingly important in acid media and for high conductivity electrolyte solutions [17]. In the presence of oxygen, part or all of the Fe(II) generated is oxidized to Fe(III) depending on the operating solution pH. Fe(III) species may be found in solution as coloidal hydrous ferric oxides (HFO) and/or onto the surface of ZVI particles forming different solid phases including $FeOOH$, Fe_3O_4 and Fe_2O_3. Hence, for ZVI/water systems the role of the corrosion products in the mechanism of arsenic removal can be dual: i- arsenic may be entrapped within the growing phase of iron oxy-hydroxydes (co-precipitation) or ii-arsenic species may be electrostatically bound to the surface of corrosion products (adsorption).

The possibility of removing simultaneously As(III) and As(V) species is one of the advantages of ZVI-based treatments operated under aerobic conditions. This seems to be possible for two reasons. On one side, the mechanism of co-precipitation involved in the removal process has been identified as an unspecific process. On the other hand, in oxygen rich waters, the reduction of dissolved oxygen yields reactive intermediates (e.g., $HO_2/O_2^{\bullet-}$, H_2O_2, $HO^{\bullet}$) which enhance the oxidation of As(III) to As(V)[18, 19], the latter species being much easier adsorbed by corrosion products.

The adsorption of As(V) onto ferric oxides is easier than that of As(III) due to the lower pK values associated to the former species. Given that the adsorption efficiency is influenced by the deprotonation ability of the sorbate and that the pK value of arsenite is higher than that of arsenate, at neutral pH values the uptake of the deprotonated form of arsenate is much more efficient than the uptake of the protonated arsenite form [20].

Taking into account the complexity of the removal mechanisms involved, a treatment system based on this technique is difficult to design. Moreover, dynamical changes in the iron bed behavior, due to the formation and transformation of corrosion products during the system lifespan, are among the main causes of efficiency reduction. Therefore, the overall system optimization requires a deep understanding of several coexisting physicochemical processes.

DESIGN OF A SMALL SCALE PLANT FOR ARSENIC REMOVAL

Taking into consideration both the information associated with the specific processes involved in ZVI/water systems and the theoretical knowledge concerning unit processes of conventional filters, a small-scale treatment plant, based on the use of Fe(0), was developed.

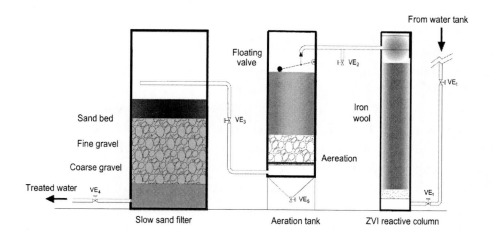

Figure 1. Small-scale ZVI-based prototype for arsenic removal.

The prototype was designed to continuously treat As-contaminated groundwater with a flow rate of 500 ml/min and with a minimum requirement of electrical energy. The objective was to assemble a system of simple operation with inexpensive and easily available materials. The system involves three stages and works with a gravity-induced flow of water. Actually, from a domestic tank water initially passes through a tubular reactor filled with ZVI. Despite several ZVI sources can be used, we have chosen iron wool because it is relatively cheap and easily available. Other authors have worked with iron powder mixed with different proportions of inert materials, granular iron or ZVI nanoparticles. During the first stage ZVI reacts with the dissolved oxygen (DO) present in the inflow water producing soluble Fe(II) species. At the same time, part of the formed Fe(II) also reacts with DO to yield Fe(III) species that form corrosion products able to co-precipitate or adsorb a fraction of the As present in the raw water inside the column. The passage of the solution to the second stage is regulated by a floating valve in the aeration tank.

During the second stage, an aeration/contact tank is used to oxidize the remaining soluble Fe(II) and to induce additional precipitation/co-precipitation processes that enhance As removal. To facilitate particle nucleation and growth, gravel is used in the tank bottom. Finally, a slow sand filter was included to remove the iron-precipitated species below the allowed drinking water Fe limit. The sand filter operates downstream and is made with a coarse gravel layer as support, a fine gravel middle layer, a sand bed and a geotextile fabric on the top. The schematic representation of the designed plant and a photograph of the complete system are shown in Figure 1.

EFFECT OF OPERATING CONDITIONS

It is widely accepted that treatment efficiency of ZVI-continuous systems mainly depends on to the nature of used Fe(0) media and on the column operating conditions [21]. In particular, the physicochemical characteristics of inflowing water and the hydraulic retention time (HRT) are among the operating conditions that most influence the performance of the overall process.

The chemistry of contaminated water determines the extent of corrosion, the iron speciation and the interactions of the contaminant with the corrosion products. Several authors have investigated the effect of different ions, pH and oxygen content on the arsenic efficiency removal [22–27].

The reactions involved on ZVI/water systems are complex and highly dependent on pH and oxygen content. High dissolved oxygen concentrations and low pH values increase the rate of ZVI oxidation to Fe(II) inside the column during the first treatment stage. Upon air saturation in the second stage, the oxidation of Fe(II) species generates large amounts of ferric oxyhydroxides, thus enhancing arsenic removal through both adsorption and precipitation processes.

It was reported that the presence of different anions impacts on the efficiency of arsenic removal in ZVI-based technologies. Some authors found that chloride, carbonate, nitrate, phosphate, sulfate and borate inhibited the arsenic removal in different extents [27]. In addition, it was reported that phosphate, silicate, and molybdate compete strongly with arsenic for sorption sites, whereas sulfate and chloride do not compete effectively [25].

The flow rate determines the hydraulic retention time (HRT) of water within each of the plant units. For the ZVI bed, HRT defines the extent of the solid-liquid contact and therefore must be long enough to achieve ZVI oxidation to Fe(II) by the dissolved oxygen naturally present in the water to be treated. The HRT in the second stage, which is operated under an air saturated atmosphere, must allow for the complete oxidation of the Fe(II) eluted from the first unit and favor the precipitation of the formed Fe(III) species as iron oxy-hydroxides.

Because the iron content in solution is a main design parameter, several studies of the corrosion processes were carried out in small scale columns filled with ZVI. The experiments aimed at quantifying the effect of variations in fluid HRT, inlet pH and ZVI loading on the reactive bed behavior. The processes were monitored by measuring pH, oxygen, Fe(II) and Fe(III) contents both at the inlet and the outlet of the reactive column. The oxygen content was measured on line by using a flux cell at the column outlet, whereas the other parameters were assessed immediately after sampling the column effluent.

Figures 2 to 4 show the effects of the HRT, the inlet pH and the ZVI loading on the generation of Fe(II) and Fe(III) species, whereas Figures 5 to 7 present the effects of the same operative variables on the rate of dissolved oxygen consumption and on the pH value at the column outlet. Briefly, results show that Fe(II) levels increase with HRT, decrease with increasing pH and do not significantly depend on ZVI loading. Despite the generation of Fe(III) species is also favored by the increase of the HRT, in contrast with the behavior observed for Fe(II) species, Fe(III) levels increase with increasing pH and strongly decrease with ZVI loading. On the other hand, dissolved oxygen consumption increases with HRT and ZVI loading, but decreases as the pH is increased. Finally, the pH values recorded at the column outlet increase with HRT and inlet pH, but show almost no dependence on the ZVI loading of the fixed bed. Noteworthy, the difference between the inlet and outlet pH values decreases as the inlet pH is increased, in agreement with previously reported trends [28].

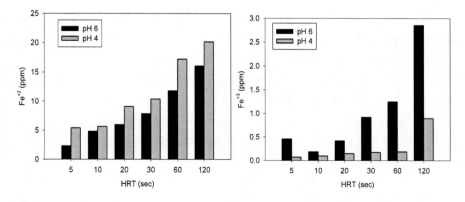

Figure 2. Effect of HRT on Fe^{+2} and Fe^{+3} species generation (m $_{ZVI}$=0.5 gr).

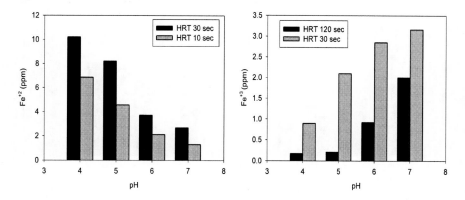

Figure 3. Effect of pH on Fe^{+2} and Fe^{+3} species generation (m_{ZVI}=0.5 gr).

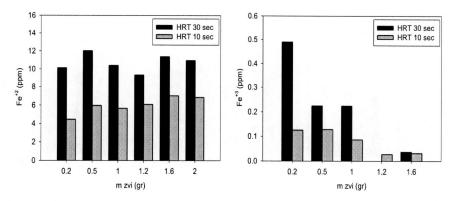

Figure 4. Effect of m_{ZVI} on Fe^{+2} and Fe^{+3} species generation (pH=4).

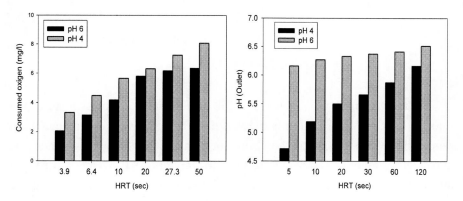

Figure 5. Effect of HRT on oxygen consumption and outlet pH.

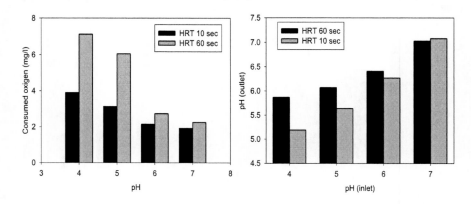

Figure 6. Effect of inlet pH on oxygen consumption and outlet pH (m_{ZVI}=0.5 gr).

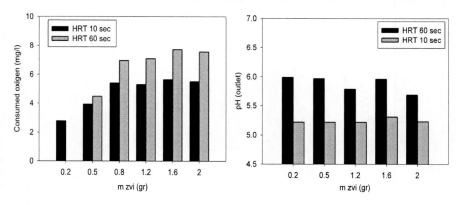

Figure 7. Effect of m_{ZVI} on oxygen consumption and outlet pH (pH_i=4).

ACCUMULATION OF CORROSION PRODUCTS AND BED CLOGGING

At near neutral pH values, typical of subsurface natural waters, the main route of ZVI oxidation is the dissolved oxygen-mediated mechanism. Under these conditions, the solubility of iron species formed is rather low and yields a layer of corrosion products that covers ZVI surface. In order to sustain ZVI oxidation, the available dissolved oxygen must migrate through the corrosion layer to access the Fe(0) surface. Therefore, the nature of the accumulated rust is of key importance. Moreover, since the soluble Fe(II) species resulting from Fe(0) oxidation must also cross the oxide layer for an efficient system operation, the processes are limited by both the decrease in the oxidant access to the surface and the diffusion of iron soluble products.

Many studies have focused on the analysis of the nature of the accumulated rust and on its temporal evolution[29, 30, 31]. Freshly formed oxides favor the diffusion inside the corrosion layer since they have a higher degree of hydration (FeOOH), are less dense, more adsorptive and more porous than aged oxides. In contrast, aged oxides (FeO, Fe_3O_4, Fe_2O_3) are usually placed close to ZVI surface and show higher dehydration degree as well as a more crystalline structure, which results in a more dense, less adsorptive and less porous layer that limits diffusion processes. For this reason, it is important to set the proper operating conditions to avoid the early formation of compact solids that lead to the inactivation of the ZVI beds. Noteworthy, it has been reported that oxidizing operating conditions yield a large array of porous oxides, while under reducing conditions the most compact crystalline structures prevail [5].

HYDRODYNAMIC BEHAVIOR OF THE ZVI FIXED BED

It is generally agreed that the accumulation of corrosion products along the column significantly affects the stability of ZVI-based systems by decreasing both the reactivity and the hydraulic conductivity of the beds. Hence, changes in column parameters such as porosity (ϕ), hydraulic conductivity (K) and residence time distribution (RTD), have a significant influence on the efficiency of the ZVI fixed beds used in water treatment systems. While the first two parameters mainly affect the volume treated per unit time and/or the associated pumping costs, the RTD directly affects the physicochemical processes involved and therefore has a direct impact on the process removal efficiency.

Hydraulic Conductivity

The assessment of the hydraulic conductivity of ZVI-based reactive columns is an important parameter for designing effective water treatment systems [21]. The falling-head and constant-head permeability methods are the most frequently used procedures for hydraulic conductivity determinations in laboratory columns and filters [32, 33, 34]. The constant head method consists on the use of a permeameter and the application of Darcy's law.

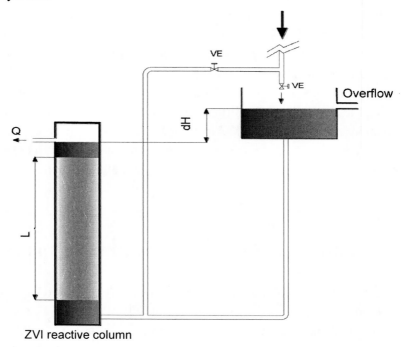

Figure 8. Constant-head permeameter coupled to the ZVI column.

This experimental law describes in a simple way the circulation of a fluid along a porous medium and states that the pore velocity (cm/min) is proportional to the hydraulic gradient (dimensionless). Laboratory determinations allow studying hydraulic conductivity variations for different column geometries or with different packing degrees. The following equation is used for the experimental evaluation of this parameter as well as its temporal evolution along the system lifespan.

$$K = \frac{Q*L}{A_t*dH}$$ (1)

where K is the hydraulic conductivity (cm/min), Q is the volumetric flow rate of the fluid through the bed (cm^3/min), L is the bed depth (cm), A_t is the total cross-sectional area of the bed (cm^2) and dH is the head loss (cm).

A laboratory test using tap water was carried out with the three-module pilot plant in order to evaluate the hydraulic conductivity of the reactive column. For this end, a constant-head permeameter was coupled in parallel to the column feed (Figure 8) and the hydraulic conductivity of the reactive bed was determined weekly.

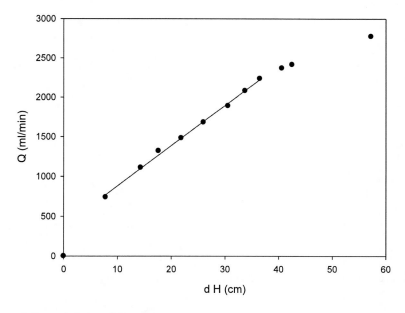

Figure 9. Darcy's law validity range.

Darcy's law fails for very low K values or high flow rates. The range of validity corresponds to a Reynolds number between 1 and 10. It has been experimentally found that, to ensure the Reynolds number does not exceed the latter validity limit, the total pressure drop in the permeameter should not be greater than 50% of the total sample length. Therefore, prior to the evaluation of the temporal evolution of K, several determinations of the initial K value were conducted for different water levels to verify the validity range of the Darcy's law for the studied system. Figure 9 shows that the law is valid for levels between 8 and 40 cm, where the flow rate varies linearly with the hydraulic load applied.

Figure 10 shows the behavior of the hydraulic conductivity against the treated volume. The obtained K values clearly decrease with the treated volume, suggesting that porosity losses and clogging effects result in an increase of the pressure drop along the bed.

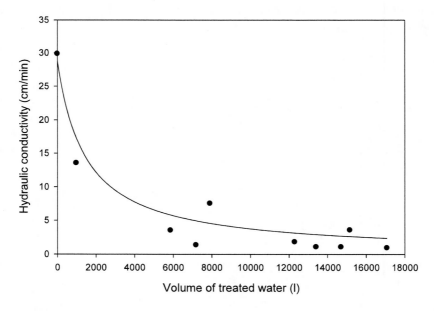

Figure 10. Temporal evolution of column hydraulic conductivity.

Taking into account that the present prototype is designed for operation with a constant pressure at the column inlet, an initial decrease in the flow rate within the initial stages of the process is expected. However, after a relatively short period, K values decrease smoothly thus allowing a rather steady operation from a hydraulic viewpoint. Noteworthy, other authors have found much more pronounced reductions in the hydraulic conductivity of ZVI beds, in some cases being of two orders of magnitude [33].

Possible causes of clogging for ZVI beds operated with real water matrices are: $CaCO_3$ precipitation, iron species precipitation, accumulation of H_2 gas in anoxic environments, growth of biofilms and accumulation of suspended particles [21], [35]. Given the operational conditions selected for our laboratory tests, the accumulation of $CaCO_3$, H_2production and biofilm growth may be neglected. Hence, the hydraulic conductivity decrease results from the accumulation of expansive iron corrosion products that lead to permeability losses [36]. It is important to note that, for field applications, operation conditions have to be properly selected in order to minimize the abovementioned causes of bed clogging.

Non-Ideal Flow

The flow-through characteristics in real reactors always show deviations from the ideal flow behavior. These are caused, among other reasons, by fluid recirculation, short-circuiting and large zones with reduced water exchange (dead zones). Deviations from the ideal behavior should be avoided as they reduce treatment efficiencies and also distort the information regarding kinetic or physicochemical parameters required for designing purposes. For the characterization of the behavior of column reactors, different strategies have been proposed. In this context, the assessment of the probability distribution function that describes the residence time of each fluid element in the reactor (RTD) is of major importance [37]. The most widely used experimental method for obtaining the RTD is the stimulus-response method, which consists in injecting a nonreactive substance (tracer) into the vessel instantaneously at a time $t = 0$ and measuring the concentration, or a tracer related property, at the reactor outlet as a function of time. The tracer should be easily detected and must not disturb the flow-through. There are different types of stimulation signals; the most frequently used being a short pulse.

In order to make the RTD independent of the tracer amount used, it is convenient to represent the RTD in a standard form (E). The E(t) curve can be determined with the values obtained at the reactor exit in the following way:

$$E(t) = \frac{C(t)}{\int_o^\infty C(t)dt} \tag{2}$$

Then, the mean residence time (3) as well as the variance of the distribution (4) can be obtained from the experimental data:

$$\bar{t} = \frac{\int_o^\infty t*C(t)dt}{\int_o^\infty C(t)dt} = \frac{\sum t_i*C_i*\Delta t_i}{\sum C_i*\Delta t_i} \tag{3}$$

$$\sigma^2 = \frac{\int_o^\infty t^2*C(t)dt}{\int_o^\infty C(t)dt} - \bar{t}^2 \cong \frac{\sum t_i^2*C_i*\Delta t_i}{\sum C_i*\Delta t_i} - \bar{t}^2 \tag{4}$$

The discrepancy between the residence time obtained as the ratio of the reactor volume to the flow rate (t_m) and the mean time obtained from equation 3 ($\bar{t}$) gives an idea of the magnitude of the deviations, which are associated with recirculation, short-circuiting and dead zones. To study and compare RTD curves from different reactor sizes or flow rates, E curve is used as a function of the dimensionless time θ.

$$\theta = \frac{t}{\bar{t}} \tag{5}$$

$$E(\theta) = \frac{C(\theta)}{\int_o^\infty C(\theta)d\theta} = E(t)*\bar{t} \tag{6}$$

The Dispersion Model

Given the design of the treatment plant developed in our laboratory, the simplest model for describing the flow pattern within the reactive bed is the ideal Plug Flow Reactor (PFRs). It is characterized by a perfect mixing in the radial dimension and no mixing in the axial dimension (i.e., no axial dispersion). In a PFR, the residence time for all elements of fluid within the reactor has to be the same.

Typically, real systems depart from this behavior. Non-ideal flow models are useful for representing the flow in real vessels. There are different kinds of models depending on whether the flow is close to a stirred tank, a plug flow or somewhere in between. For tubular reactors, two models exist to explain deviations from the plug flow: the axial dispersion model (ADM) and the tanks-in-series model (TSM). These models apply to turbulent flow in pipes, laminar flow in very long tubes, flow in packed beds, etc. [38]. For the analysis of the behavior of the ZVI column tested, the ADM model is much more appropriated than the TSM model.

The ADM is a plug flow model with an axial dispersion of the material, governed by an equation analogue to the Fick's law of diffusion, superimposed on the flow. Assuming the validity of this model, the material balance for a tracer is given by [39]:

$$\frac{\partial c}{\partial t} = -u\frac{\partial c}{\partial z} + D\frac{\partial^2 c}{\partial z^2} \tag{7}$$

where D is the axial dispersion coefficient, a parameter that is different from the molecular diffusion coefficient but describes the evolution of the solute distribution along the longitudinal axis of the column. Reordering parameters, a dimensionless form the basic differential equation is obtained:

$$\frac{\partial c}{\partial \theta} = \left(\frac{D}{uL}\right)\frac{\partial^2 c}{\partial Z} - \frac{\partial c}{\partial Z} \tag{8}$$

with $\theta = \frac{t}{\bar{t}}$ and $Z = \frac{z}{L}$. The dimensionless parameter $\left(\frac{D}{uL}\right)$ is called the vessel dispersion number and quantifies the extent of axial dispersion. Values of this parameter tending to zero represent small deviations from the plug-flow and high values represent large dispersion so the system resembles a stirred tank.

Dispersion Degrees and Boundary Conditions

The ADM equation can be fitted for different dispersion degrees and boundary conditions. If an ideal tracer pulse is injected, the dispersion inside the vessel causes the broadening of the RTD curve. For low vessel dispersion degrees ($\frac{D}{uL}$< 0.01), the curve does not change its form and the resolution of equation 8 yields a family of Gaussian curves. For large deviations from the plug-flow ($\frac{D}{uL}$> 0.01) the RTD broadens and changes its shape resulting in rather asymmetric curves. Moreover, the mathematical solutions of the latter differential equation depend on the boundary conditions, so the behavior of the tracer just at the column inlet and exit also determine the shape of $E(\theta)$ curves.

Figure 11 shows the shapes of $E(\theta)$ curves for different dispersion degrees. The profiles show that as the vessel dispersion number increases, the DTR is substantially broadened.

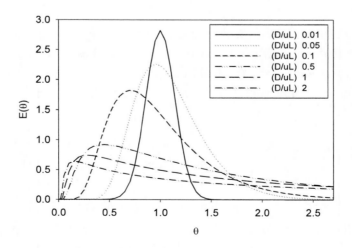

Figure 11. E(θ) curves for different dispersion degrees.

Table 1. Analytical expression obtained for different boundary conditions

	Analytical expression
Small extents of dispersion ($\frac{D}{uL}<0.01$)	$$E(\theta)=\sqrt{\frac{1}{4\frac{D}{uL}\pi}}\exp\left[-\frac{(1-\theta)^2}{4\frac{D}{uL}}\right]$$ $$\bar{\theta}_E=1\quad\sigma_E^2=2\frac{D}{uL}$$
Large deviation from plug flow($\frac{D}{uL}>0.01$)	
Open vessel	$$E(\theta)=\sqrt{\frac{1}{4\frac{D}{uL}\theta}}\exp\left[-\frac{(1-\theta)^2}{4\frac{D}{uL}\theta}\right]$$ $$\bar{\theta}_{E\infty}=1-2\frac{D}{uL}\quad\sigma_{E\infty}^2=2\frac{D}{uL}+8\left(\frac{D}{uL}\right)^2$$
Closed vessel	No analytical solution for $E(\theta)$ $$\bar{\theta}_E=1\quad\sigma_E^2=2\frac{D}{uL}-2\left(\frac{D}{uL}\right)^2\left[1-e^{-\frac{uL}{D}}\right]$$
Closed (with dispersion) – open vessel	$$E(\theta)=\sqrt{\frac{1}{\pi\frac{D}{uL}\theta}}\exp\left[-\frac{1}{4\frac{D}{uL}\theta}\right]$$ $$-\frac{1}{2\frac{D}{uL}}\exp\left[\frac{uL}{D}\right]\text{erfc}\left[\sqrt{\frac{1}{\theta\frac{D}{uL}}}\frac{1+\theta}{2}\right]$$
Closed (without dispersion) – open vessel	$$E(\theta)=\sqrt{\frac{1}{4\pi\frac{D}{uL}\theta^3}}\exp\left[-\frac{(1-\theta)^2}{4\frac{D}{uL}\theta}\right]$$

Boundary conditions significantly affect the ADM solution and include: *i)* closed boundary conditions, where it is assumed that the dispersion only occurs inside the vessel, and the fluid enters and exits as a plug-flow; *ii)* open boundary conditions, which consider that there are dispersions before the tracer enters and after it leaves the vessel; and *iii)* combinations between both. Assuming a δ-Dirac function for the injected pulse (perfect pulse), the signal at the system output can be obtained from the solution of equation 8 using different boundary conditions. The analytical expressions are summarized in Table 1 [40]. In order to characterize the behavior of the reactive column of the developed prototype, the stimulus-response method was applied and the RTD was analyzed using IK as a tracer.

The RTD curve was experimentally determined by measuring the fluid conductivity at the system exit. During one month of operation, measurements were weekly performed for obtaining the parameters that characterize the temporal evolution of the hydraulic behavior of the reactive bed. The best description of the recorded profiles was obtained by using the expression for large deviations with boundary conditions corresponding to a closed (without dispersion)-open vessel, which agrees with the experimental situation and the conductivity sampling method. Figure 12 shows the time evolution of the RTD obtained by fitting the experimental data with the equation shown in the last row of Table 1. It can be clearly seen that by increasing the operation time of the fixed bed, the axial dispersion of the treated fluid increases. This behavior may be ascribed to the appearance of preferential flow pathways and back-mixing owing to the accumulation of corrosion products inside the column.

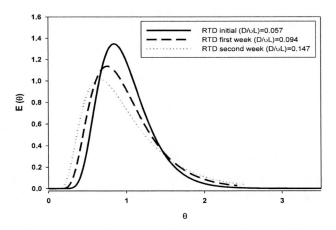

Figure 12. Temporal evolution of RTDs, obtained by fitting the experimental data to a closed (without dispersion)-open vessel model.

FIELD TESTS

According to the results obtained for small-scale columns and after the characterization of the hydraulic behavior of the system, a ZVI-based pilot-scale water treatment plant was constructed for field experiments. The main objective of this test was to verify the system efficiency for obtaining arsenic-free drinking water from a real source of groundwater naturally contaminated with arsenic. A village in Buenos Aires province was selected as test point, where groundwater from a non operative well with high arsenic content was gently provided for the field investigation.

Based on the design already proposed and the characteristics of the installation point, the materials necessary to construct a prototype for the treatment of a flow rate of 700 L/day were selected. The objective was to obtain an easy to assemble equipment with a minimum amount of energy requirement and simple operation for non-specialized users. An operating lifespan of 30 days was estimated for an initial ZVI loading of 500 gr.

PVC was chosen as construction material for the column and second module. The filter was constructed using a 60 L standard plastic container filled with filtering materials of different grain size. The prototype also included an automatic pH regulation system incorporated into the water supply.

Before the field test, some preliminary experiments were carried out in our laboratory in order to detect failures and perform the necessary modifications. Once verified the correct functioning of the prototype in the laboratory, we proceeded with its installation at the selected village and the first field test was started.

The local operators performed daily tasks including: flow rate control, pH registration and prototype supervision. In addition, samples of the treated water were periodically taken for analysis (i.e., three times per week). In situ determination of the arsenic content was performed by the local operators with the aid of test strips, whereas Fe(II), Fe(III), Fe_{Total} and As contents were determined at the University's laboratory. Fe(II) concentrations were assessed by the modified o-phenanthroline colorimetric method. Fe(III) levels were also monitored by a colorimetric method, in this case by a red complex formation between Fe(III) and KSCN. Arsenic concentration in water samples was analyzed by using the standardized silver diethyldithiocarbamate method (AgDDTC).

During the first field test some minor operational problems, related to the excessive sludge generation, were detected and quickly solved. In spite of this, removal rates of 90-95% were reached over a period of 20 days, obtaining a treated water with less than 10 ppb of As. Given the operational problems found during the first test, the sand filter size was increased in order to reduce the cleaning frequency of the last module. The final prototype used is shown in Figure 13.

Figure 13.Prototypes designed for the field tests.

For the second field test, two treatment cycles were studied by renewing the iron wool after exhaustion during the first stage. The same protocol of plant monitoring was carried out, but with the addition of three extra samples (i.e., one corresponding to the untreated water and the other two corresponding to the treated waters obtained during each cycle). To ensure a complete independence from the staff of the University of La Plata, the extra samples were sent to a private laboratory for a complete physico-chemical analysis. It is worth mentioning that the results of this second test were excellent. Removal rates of 95-100% were obtained for both periods of 28 days each (see Table 2 below). The results of the external laboratory also verified the process efficiency showing As concentrations for the two treated samples below the detection limit of the analysis technique (i.e., 3 ppb) and maintaining all the physico-chemical parameters within the established limits for drinking water by the Argentinian Food Code.

Table 2. Results obtained in the second field test*

Day (1st cycle)	Fe$_{Total}$ (ppm)	As (ppm) field	As (ppm) laboratory	Day (2nd cycle)	Fe$_{Total}$ (ppm)	As (ppm) field	As (ppm) laboratory
Start	-	0.00	-	Start	0.13	0.00	0.03
2	0.02	0.00	0.00	2	0.09	0.00	0.01
4	0.04	0.00	0.00	6	0.19	0.00	0.00
6	0.00	0.00	0.00	7	0.05	0.00	0.00
7	0.01	0.00	0.01	11	0.02	0.00	0.00
9	0.22	0.00	0.00	12	0.02	0.00	0.00
11	0.00	0.00	0.00	14	0.01	0.00	0.00
12	0.05	0.00	0.02	16	0.01	External analysis	
14	0.00	0.00	0.00	17	-	0.00	0.00
16	0.22	0.00	0.01	20	-	0.00	0.00
17	0.01	0.00	0.00	27	-	0.01	-
18	External analysis			28		0.03	End of test
19	0.00	0.00	0.00				
21	0.00	0.00	0.00				
22	0.00	0.00	0.01				
24	0.00	0.00	0.00				
26	0.00	0.00	0.00				
27	0.00	0.00	0.00				
28	0.00	0.01	0.02				
29	0.00	0.01	End of test				

*Argentinean Food Code Limits: Fe$_{Total}$=0.30 ppm As=0.01 ppm.

CONCLUSION

The high content of arsenic in groundwater is an important problem for small communities and rural areas in Argentina. The results of the present chapter demonstrate that water treatment plants based on the use of ZVI fixed beds may provide a technical solution essential for the population affected The developed prototype may be easily assembled using relatively inexpensive and widely available materials.

While the detailed chemical mechanisms of the ZVI process are complex and the temporal evolution of the corrosion products affects the system efficiency, the characterization of operative parameters allows an efficient design of the drinking water treatment plants. Critical design parameters of the ZVI system are the hydraulic retention time (HRT) within the ZVI column and the pH at the column inlet. The HRT within the column should be long enough for allowing an efficient ZVI oxidation by the dissolved oxygen naturally present in the water to be treated. High pH values give very low corrosion rates, whereas low pH values generate excessive amounts of sludge. The excess of corrosion products can reduce the bed lifespan since its accumulation along the column decreases the hydraulic conductivity (K) and increases the dispersion degree of the plug-like flow. Since ZVI columns leach significant quantities of iron, a precipitation tank and a sand filter should be placed after the iron bed for obtaining drinking water of adequate quality.

Laboratory and field tests demonstrated that the three-module ZVI-based system is capable of efficiently removing arsenic from groundwater to levels below the WHO guideline of 10 ppb. The developed plant can be operated by non-specialized personal, requires minimum maintenance, is designed to continuously deliver safe drinking water during four weeks with an average flow rate of at least 700 L per day and uses only 500 gr of iron wool per service cycle.

REFERENCES

[1]   A. B. Cundy, L. Hopkinson, and R. L. D. Whitby, "Use of iron-based technologies in contaminated land and groundwater remediation: A review," *Sci. Total Environ.,* vol. 400, no. 1–3, pp. 42–51, 2008.

[2]   L. B. Young and H. H. Harvey, "The Relative Importance of Manganese and Iron-Oxides and Organic-Matter in the Sorption of

Trace-Metals by Surficial Lake-Sediments," *Geochim. Cosmochim. Acta,* vol. 56, pp. 1175–1186, 1992.

[3] L. C. Roberts, S. J. Hug, T. Ruettimann, M. Billah, A. W. Khan, and M. T. Rahman, "Arsenic Removal with Iron(II) and Iron(III) in Waters with High Silicate and Phosphate Concentrations," *Environ. Sci. Technol.,* vol. 38, no. 1, pp. 307–315, 2004.

[4] X. Han, J. Song, Y. Li, S. Jia, W. Wang, and F. Huang, "As(III) removal and speciation of Fe (Oxyhydr)oxides during simultaneous oxidation of As(III) and Fe(II)," *Chemosphere,* vol. 147, pp. 337–344, 2016.

[5] C. Noubactep, "Processes of contaminant removal in 'Fe0 -H2O' systems revisited: the Importance of co-precipitation," *Environ. Sci.,* pp. 9–13, 2007.

[6] L. Gui, Y. Q. Yang, S. Jeen, R. W. Gillham, D. W. Blowes, and C. Noubactep, "Comment on '' Reduction of chromate by granular iron in the presence," *Appl. Geochemistry,* vol. 24, no. 11, pp. 2206–2207, 2009.

[7] C. Noubactep, "A critical review on the process of contaminant removal in Fe0-H2O systems," *Environ. Technol.,* no. June 2014, pp. 909–920, 2008.

[8] R. S. Burkel and R. C. Stoll, "Naturally occurring arsenic in sandstone aquifer water supply wells of North Eastern Wisconsin," *Gr. Water Monit. Remediat.,* vol. 19, no. 2, pp. 114–121, 1999.

[9] World Health Organization, *Guide lines for drinking-water qualitity recomendations,* Second. Geneva, 1993.

[10] J. Bundschuh, M. I. Litter, F. Parvez, G. Román-ross, H. B. Nicolli, J. Jean, C. Liu, D. López, M. A. Armienta, L. R. G. Guilherme, A. Gomez, L. Cornejo, L. Cumbal, R. Toujaguez, and C. Rica, "One century of arsenic exposure in Latin America : A review of history and occurrence from 14 countries," *Sci. Total Environ.,* vol. 429, pp. 2–35, 2012.

[11] H. B. Nicolli, J. M. Suriano, M. A. G. Peral, L. H. Ferpozzi, and O. A. Baleani, "Groundwater Contamination with Arsenic and Other Trace Elements in an Area of the Pampa, Province of Cordoba,

Argentina," *Environ. Geol. and Water Sci.,* vol. 14, no. 1, pp. 3–16, 1989.

[12] P. L. Smedley, H. B. Nicolli, D. M. J. Macdonald, A. J. Barros, and J. O. Tullio, "Hydrogeochemistry of arsenic and other inorganic constituents in groundwaters from La Pampa, Argentina," *Appl. Geochemistry,* vol. 17, no. 3, pp. 259–284, 2002.

[13] D. Mohan and C. U. Pittman, "Arsenic removal from water/wastewater using adsorbents-A critical review," *J. Hazard. Mater.,* vol. 142, no. 1–2, pp. 1–53, 2007.

[14] S. F. O'Hannesin and R. W. Gillham, "Long-Term Performance of an in situ 'Iron Wall' for Remediation of VOCs," *Ground Water,* vol. 36, pp. 164–170, 1998.

[15] D. H. Phillips, T. Van Nooten, L. Bastiaens, M. I. Russell, K. Dickson, S. Plant, and R. M. Kalin, "Ten Year Performance Evaluation of a Field-Scale Fe0 Permeable Reactive Barrier Installed to Remediate TCE-Contaminated Groundwater," *Environ. Sci. Technol.,* vol. 44, pp. 3861–3869, 2010.

[16] R. T. Wilkin, S. D. Acree, R. R. Ross, R. W. Puls, T. R. Lee, and L. L. Woods, "Fifteen-year assessment of a permeable reactive barrier for treatment of chromate and trichloroethylene in groundwater," *Sci. Total Environ.,* vol. 468–469, pp. 186–194, 2014.

[17] U. R. Evans, "Use of soluble inhibitors," *Ind. Eng. Chem.,* no. 8, pp. 703–705, 1945.

[18] M. J. López-Muñoz, A. Arencibia, Y. Segura, and J. M. Raez, "Removal of As(III) from aqueous solutions through simultaneous photocatalytic oxidation and adsorption by TiO_2 and zero-valent iron," *Catal. Today,* vol. 280, pp. 149–154, 2017.

[19] O. X. Leupin and S. J. Hug, "Oxidation and removal of arsenic (III) from aerated groundwater by filtration through sand and zero-valent iron," *Water Res.,* vol. 39, pp. 1729–1740, 2005.

[20] K. Banerjee, R. P. Helwick, and S. Gupta, "A treatment process for removal of mixed inorganic and organic arsenic species from groundwater," *Environ. Prog.,* vol. 18, no. 4, pp. 280 – 284, 1999.

[21] R. Domga, F. Togue-kamga, C. Noubactep, and J. Tchatchueng, "Discussing porosity loss of Fe0 packed water filters at ground level," *Chem. Eng. J.,* vol. 263, pp. 127–134, 2015.

[22] J. M. Triszcz, A. Porta, and F. S. García, "Effect of operating conditions on iron corrosion rates in zero-valent iron systems for arsenic removal," *Chem. Eng. J.,* vol. 150, pp. 431–439, 2009.

[23] S. Bang, M. D. Johnson, G. P. Korfiatis, and X. Meng, "Chemical reactions between arsenic and zero-valent iron in water," *Water Res.,* vol. 39, pp. 763–770, 2005.

[24] S. Bang, G. P. Korfiatis, and X. Meng, "Removal of arsenic from water by zero-valent iron," *J. Hazard. Mater.,* vol. 121, no. 1–3, pp. 61–67, 2005.

[25] C. Su and R. W. Puls, "Arsenate and Arsenite Removal by Zerovalent Iron : Effects of Phosphate, Silicate, Carbonate, Borate, Sulfate, Chromate, Molybdate, and Nitrate, Relative to Chloride," *Environ. Sci. Technol.,* vol. 35, no. 22, pp. 4562–4568, 2001.

[26] H. Sun, L. Wang, R. Zhang, J. Sui, and G. Xu, "Treatment of groundwater polluted by arsenic compounds by zero valent iron," *J.Hazard. Mater.,* vol. 129, no. 1–3, pp. 297–303, 2006.

[27] M. Biterna, L. Antonoglou, E. Lazou, and D. Voutsa, "Arsenite removal from waters by zero valent iron: Batch and column tests," *Chemosphere,* vol. 78, no. 1, pp. 7–12, 2010.

[28] S. J. Hug, "pH Dependence of Fenton Reagent Generation and As (III) Oxidation and Removal by Corrosion of Zero Valent Iron in Aerated Water," *Environ. Sci. Technol.,* vol. 42, no. 19, pp. 7424–7430, 2008.

[29] J. suk O, S. W. Jeen, R. W. Gillham, and L. Gui, "Effects of initial iron corrosion rate on long-term performance of iron permeable reactive barriers: column experiments and numerical simulation.," *J. Contam. Hydrol.,* vol. 103, no. 3–4, pp. 145–56, 2009.

[30] J. Dries, L. Bastiaens, D. Springael, S. Kuypers, S. N. Agathos, and L. Diels, "Effect of humic acids on heavy metal removal by zero-valent iron in batch and continuous flow column systems," *Water Res.,* vol. 39, no. 15, pp. 3531–3540, 2005.

[31] C. Noubactep, "An analysis of the evolution of reactive species in Fe0/H2O systems," *J. Hazard. Mater.*, vol. 168, no. 2–3, pp. 1626–1631, 2009.

[32] N. Moraci and P. S. Calabró, "Heavy metals removal and hydraulic performance in zero-valent iron/pumice permeable reactive barriers," *J. Environ. Manage.*, vol. 91, no. 11, pp. 2336–2341, 2010.

[33] D. N. H. Beach, J. E. McCray, K. S. Lowe, and R. L. Siegrist, "Temporal changes in hydraulic conductivity of sand porous media biofilters during wastewater infiltration due to biomat formation," *J. Hydrol.*, vol. 311, no. 1–4, pp. 230–243, 2005.

[34] D. Mishra and J. Farrell, "Evaluation of mixed valent iron oxides as reactive adsorbents for arsenic removal," *Environ. Sci. Technol.*, vol. 39, no. 24, pp. 9689–9694, 2005.

[35] P. D. Mackenzie, D. P. Horney, and T. M. Sivavec, "Mineral precipitation and porosity losses in granular iron columns," *J. Hazard. Mater.*, vol. 68, no. 1–2, pp. 1–17, 1999.

[36] B. D. Btatkeu-k, H. Olvera-vargas, J. B. Tchatchueng, C. Noubactep, and S. Caré, "Determining the optimum Fe0 ratio for sustainable granular Fe0 /sand water filters," *Chem. Eng. J.*, vol. 247, pp. 265–274, 2014.

[37] O. Levenspiel, "Basics of Non-Ideal flow," in *Chemical Reaction Engineering, Third Edit.*, John Wiley & Sons, Inc., 1999, pp. 257–282.

[38] O. Levenspiel, "The Dispersion Model," in *Chemical Reaction Engineering, Third Edit.*, John Wiley & Sons, Inc., 1999, pp. 293–320.

[39] A. Kołodziej, M. Jaroszy, H. Schoenmakers, K. Althaus, M. Kloeker, E. Geißler, and C. Ubler, "Dynamic tracer study of column packings for catalytic distillation," vol. 44, pp. 661–670, 2005.

[40] A. N. Colli, "Estudio teórico y experimental de las desviaciones de la idealidad en reactores electroquímicos," [Theoretical and experimental study of the deviations of ideality in electrochemical reactors], Universidad Nacional del Litoral, 2013.

BIOGRAPHICAL SKETCHES

Fernando S. García Einschlag

Affiliation: Instituto de Investigaciones Fisicoquímica Teóricas y Aplicadas (INIFTA), CCT-La Plata-CONICET, Departamento de Química, Facultad de Ciencias Exactas, UNLP, Buenos Aires, Argentina.

Education: Biochemist. PhD in Exact Sciences

Business Address: Diag 113 y 64, La Plata, Buenos Aires. Argentina

Research and Professional Experience: Physical Chemistry. Water treatment and wastewater treatment

Professional Appointments: Adjoint Professor of Physical Chemistry, UNLP. Independent Researcher of CONICET

Honors: 2016 Innovation Award of the University of La Plata (development of a treatment plan for arsenic contaminated waters).

Publications from the Last 3 Years:

- L. Santos-Juanes; F.S. García Einschlag; A.M. Amat; A. Arques.Combining ZVI reduction with photo-Fenton process for the removal of persistent pollutants. *Chemical Engineering Journal*; Lugar: Amsterdam; Año: 2016.
- Alejandra Saavedra Moncada; Fernando S. García Einschlag; Eduardo D. Prieto; Gustavo T. Ruiz; Alexander G. Lappin*; Guillermo J. Ferraudi*; Ezequiel Wolcan*. Photophysical properties of -Re(I)(CO)3(phen) pendants grafted to a poly-4-vinylpyridine backbone. A correlation between photophysical properties and morphological changes of the backbone. *Journal of*

Photochemistry and Photobiology A-Chemistry; Lugar: Amsterdam; Año: 2016 vol. 321 p. 284 – 296.

- Daniela A. Nichela; Jorge A. Donadelli; Bruno F. Caram; Menana Haddou; Felipe J. Rodriguez Nieto; Esther Oliveros; Fernando S. García Einschlag*. Iron cycling during the autocatalytic decomposition of benzoic acid derivatives by Fenton-like and photo-Fenton techniques. *Applied Catalysis B-Environmental*; Lugar: Amsterdam; Año: 2015 vol. 170 p. 312 – 321.

- M. Micaela Gonzalez; M. Paula Denofrio; Fernando S. García Einschlag; Carlos A. Franca; Reinaldo Pis Diez; Rosa Erra-Balsells; Franco M. Cabrerizo. Determining the Molecular Basis for the pH-dependent Interaction among 2′-deoxynucleotides and 9H-pyrido[3,4-b]indole in its ground and electronic excited states. *Physical Chemistry Chemical Physics*; Lugar: Cambridge; Año: 2014 vol. 16 p. 16547 – 16562.

- Fernando S. García Einschlag; André M. Braun; Esther Oliveros*. Fundamentals and Applications of the Photo-Fenton Process to Water Treatment (Chapter 247). *Handbook of Environmental Chemistry: Environmental Photochemistry Part III*. Lugar: Berlín - Heidelberg; Año: 2015; p. 301 – 342.

Eliana Berardozzi

Affiliation: Instituto de Investigaciones Fisicoquímica Teóricas y Aplicadas (INIFTA), CCT-La Plata-CONICET, Departamento de Química, Facultad de Ciencias Exactas, UNLP, Buenos Aires, Argentina. Departamento de Hidráulica, Facultad de Ingeniería, UNLP.

Education: Chemical Engineer

Business Address: Diag 113 y 64, La Plata, Buenos Aires. Argentina

Research and Professional Experience: Physical Chemistry. Water treatments.

Professional Appointments: PhD Student (CONICET) and assistant lecturer in Chemical Engineering (UNLP).

Honors: 2016 Innovation Award of the University of La Plata (development of a treatment plan for arsenic contaminated waters)

Publications from the Last 3 Years:

- Eliana Berardozzi; Gabriela Ortigoza Medina; Cecilia Lucino; Fernando S. García Einschlag. Planta piloto para la remoción de arsénico en agua: prueba de campo. VI Congreso Internacional sobre gestión y tratamiento integral del agua; Año: 2016. [Pilot plant for the removal of arsenic in water: field test. VI *International Congress on water management and integral treatment*; Year: 2016.]
- Eliana Berardozzi; Andrés J. Donadelli; Fernando S. García Einschlag. Influencia de las condiciones operativas sobre la producción de Fe(II) en columnas rellenas con hierro cero valente: análisis de las superficies de respuesta. XIX Congreso Argentino de Fisicoquímica y Química Inorgánica. Año: 2015. [Influence of the operating conditions on Fe (II) production in columns filled with zero-valent iron: analysis of the response surfaces. XIX *Argentine Congress of Physical Chemistry and Inorganic Chemistry*. Year: 2015]
- Eliana Berardozzi; Gabriela Ortigoza Medina; Cecilia Lucino; Fernando S. García Einschlag. Remoción de As en columnas con ZVI. Efecto de las variables operativas. Argentina y Ambiente 2015. Año: 2015.[Removal of As in columns with ZVI. Effect of operating variables. *Argentina and Environment* 2015. Year: 2015.]

In: Arsenic
Editor: Ratko Knežević

ISBN: 978-1-53612-461-3
© 2017 Nova Science Publishers, Inc.

Chapter 4

TOXICOGENESIS AND METABOLISM OF ARSENIC IN RICE AND WHEAT PLANTS WITH PROBABLE MITIGATION STRATEGIES

Arnab Majumdar and Sutapa Bose, PhD*
Earth and Environmental Science Research Laboratory,
Department of Earth Sciences, Indian Institute of Science Education
and Research Kolkata, West Bengal, India

ABSTRACT

The metalloid arsenic (As) is a natural constituent, contaminating the agro-ecosystem and gets increased along with the expansion of human health risks due to anthropogenic pollution. Rice and wheat are the staple cereal crops that are being cultivated in arsenic contaminated fields globally. From the very seedling stage to the final harvesting of crops, the effect of arsenic is prominently resulting in declination of root-shoot length and enzymatic expressivity in both wheat and paddy plants. Antioxidant enzymatic responses viz. superoxide dismutase (SOD), ascorbate peroxidase (APX), glutathione reductase (GR), and a few, get

---

* Address correspondence to: sutapa.bose@iiserkol.ac.in.

compromised immensely due to the excess bioavailability of arsenic to the plant system. Speciation of arsenic in its As(III) and As(V) forms are predominant respectively in aqueous and aerobic soil profile entangling with other elements in the system for the transportation inside the plant. Plants like rice and wheat tend to accumulate a greater percentage of arsenite [As(III)] due to its higher mobility and solubility that let the access of transportation through plant transporter molecules known as nodulin-like intrinsic proteins (NIPs) or aquaporins and Lsi transporters. Arsenate [AS(V)] commonly compete with phosphate molecule for the passage through Pht molecules by having a similar molecular structure. Apart from the inorganic species, organic forms of arsenic might get passed through the plant uptake system but less effectively. Though wheat and paddy plants are not hyperaccumulators of arsenic, some of the wetland plants namely, *Pteris* spp., *Eichhornia* spp., Agrostis spp., can be helpful in course of phytoremediation of arsenic before rice or wheat cultivation on contaminated fields. Phytochelatins are plant derived compounds, a derivative of glutathione that might form chelates with arsenic and can be a mode of plant detoxification of excess arsenic accumulation inside the plant tissue system. This glimpse of previous literature and current studies will enlighten the toxicogenesis of arsenic in wheat and paddy plant with subsequent internal metabolism and remediation strategies furthermore.

Keywords: anthropogenic pollution, arsenic toxicogenesis, antioxidant enzymes, transporter proteins, hyperaccumulator, phytochelatins

1. INTRODUCTION

Arsenic, despite being a natural element among earth's compositions, its presence in soil-sediments to groundwater and plant system to animal tissues beyond the expectable limit, contamination possess the threat to a natural sustainability. Arsenic (As) remains in our system mostly in two available form; arsenate [As(V)] and arsenite [As(III)], although there are three other different oxidative states of arsenic is present in the environment, namely as arsenic [As(0)], arsine [As(-III)], organo-arsenic compounds like monomethyl arsenic (MMA), dimethyl arsenic (DMA), trimethylarsenic oxide (TMAO) and likely so (Huang and Matzner, 2006). Several studies have been made to find out the toxicity of arsenic in the soil to plant system and subsequently in respective consumers which appeal for the mass awareness about this anthropogenic disaster (Meharg, 2004; Sahoo and Mukherjee, 2014). From geogenic perspective, arsenic gets released to the surface soil through several processes like volcanic eruptions, landslides, or severe man-made activities that trigger the deeper earth elements to come out. On the other hand, continuous and inappropriate digging of bore wells and shallow tube wells resulted in the groundwater aquifers to be contaminated and once a contaminated groundwater aquifer is quite stringent to get the system contamination free again. In such a manner, several years of exploitation of groundwater, contaminated with arsenic, agricultural soil and subjective crop system becomes contaminated also with such practices. Toxicity of arsenic in rice and wheat plants reveals the two scenarios clear; the transportation of arsenic and its subsequent metabolism is not qualified by the plant system and secondly, its adverse effect on the consumers.

Transfer of arsenic from soil to the plant root system and gradually to the upper ground parts depend on the presence of other chemical constituents in the soil with relative soil physico-chemical characteristics (Shrivastava et al., 2014; Barla et al., 2017). Terrestrial plants are able to take up a lesser amount of arsenic and other toxic metal(loids) compared to any wetland plant species. Arsenic is mostly present in soil environment as arsenate [As(V)] and its related conjugates, whereas, in an aqueous phase, arsenic predominates as arsenite [As(III)] which is much soluble to the aqueous phase. Structurally, arsenate is competitive to the phosphate whereas arsenite shows similarities to the water molecules that play the tricks of the arsenic transportation in most of the wetland plant species compared to the terrestrial vegetation.

2. ARSENIC TRANSPORTATION PRIOR TO ITS TOXICITY

Arsenic transportation occurs from the contaminated soil to the plant root system where Fe plaque predominates and this allows the arsenic uptake kinetic to be altered (Liu et al., 2006). Due to the presence of Fe plaque with its various oxides/hydroxides in aerobic soils, having a strong affinity to the arsenate, rice root cells sometimes absorb partial arsenate concentration along with some arsenite uptake. As(V) is an analog of phosphate and that allows the arsenate molecules to be transported through the phosphate transporter of rice plants and studies also suggests that presence of high concentration of arsenate induces phosphate signaling molecules and even direct to misleading sensing of phosphate. On the basis of plant species variation, transporters of phosphate might show a different affinity towards the arsenate like arsenate hyperaccumulator fern *Pteris vittata* compared to other terrestrial non-hyperaccumulator plants (Poynton et al., 2004). Species vary with phosphate transporters that commonly involve the Phosphate transporter 1 (Pht 1) family with more than 100 transporter genes which are more often present in greater number around the root vascular cells (Bucher, 2007). Earlier studies have reported that *Arabidopsis thaliana* showed up with two phosphate transporter (Pht 1;1 and Pht 1;4) with both phosphate rich and arsenate rich system that transport arsenate in considerable amount through these proteins in wild type of the model organism. But in the double mutated (Pht 1;1Δ4Δ) of *A. thaliana*, the arsenate transportation and resistance was much higher compared to the wild organism (Shin et al., 2004). This emphasizes on the relation between arsenate and phosphate interaction at the cellular level in plant root system. Arsenite, on the other hand, is predominant in reducing environments like flooded wetland system where a minor change in redox state reduces arsenate to arsenite. Arsenous acid [As(OH)$_3$] resemblance with silica hydroxide [Si(OH)$_4$] and this led to the competition between silica and As(III) for the transportation from an exoplasmic environment to the root cell cytosol via specific silicon transporters (Bhattacharjee and Rosen, 2007). Transporters like Lsi1 passes arsenite from surrounding environments to the root cells whereas some other transporters like Lsi2 mediate this process from root cells or other tissues to xylem parenchyma (Ma et al., 2006). Nodulin-26-like intrinsic proteins (NIP) is a group of transmembrane transporter proteins that allows water, glycerol, glucose and some smaller similar types of molecules to pass through including arsenite. These NIP proteins consist diversified members that are present in almost every category of plant species and identified as similar generic names with different numeric numbers to denote the origin of those proteins from respective genes. *AtNIP1;1, AtNIP1;2, AtNIP5;1, AtNIP7;1, LjNIP5;1, LjNIP6;1,OsNIP1;1, OsNIP3;1* are some of the NIP protein transporters identified from *Arabidopsis thaliana, Oryza sativa*, and *Lotus japonicas* respectively and divided in three group of NIP families (Zhao, McGrath and Meharg, 2010). Interestingly, detail findings on organoarsenic species like DMAA, MMAA and their transport mechanisms were not established except a few that suggests that the transportation might take place through the same silicon transporter as well as phosphate transporter, although the chance of inorganic arsenic transportation is much higher than any organoarsenic species.

3. TOXICOGENESIS AND ARSENIC STRESS IN SEEDLINGS

The presence of arsenic or some other toxic elements in surrounding growth medium of cereals, plants take up the primary available form of elements through root epidermis cells and consequently transport to the upper ground parts with a notable reduction in physiological imbalance. Arsenic and other heavy metal accumulation at the rhizospheric zone allows translocation of both As(III) and As(V) along with some DMAA and MMAA which on the other hand volatilized partially after their methylation process (Zhao, McGrath and Meharg, 2010). This results in less storage of arsenic inside the plant body parts, although in seedlings of rice and wheat shows considerable toxicity in the presence of high concentration of arsenic. Plant enzymes that manage environmental stresses are the major parameter of observation for the arsenic toxicogenesis in seedlings. To sustain in adverse conditions, plants have evolved with diversified metabolic precursor molecules that can be triggered up to its functionalized state and cope up with the environmental stress. This includes specifically, the product of stress tolerance genes or enzymes, connected through a complex chemical signal transduction pathway (Hirayama and Shinozaki, 2010). Arsenic or any other toxic metals that creates toxicity to the plant system above their natural tolerance capacity, imparts also some Reactive oxygen species (ROS), including superoxide radical ($O_2{}^{\cdot-}$), singlet oxygen ($_1O^2$), hydroxyl radical ($OH^{\cdot}$), hydrogen peroxide (H_2O_2) etc., coming out from the incomplete reduction or excitation of molecular oxygen in aerobic condition (You and Chan, 2015). Sustained oxygenic species in such stressed conditions become deleterious to the plant cell proteins, lipids and even nuclear materials leading to the extracellular or intracellular damages leading to the death of plant cells. To bypass this hazardous effect, plants are evolved with a series of ROS scavenging antioxidant enzymes. Superoxide dismutase (SOD), catalase (CAT), glutathione S-transferase (GST), glutathione reductase (GR), glutathione peroxidase (GPX), ascorbate peroxidase (APX), dehydroascorbate reductase (DHAR), monodehydroascorbate reductase (MDHAR), and peroxiredoxin (PRX) is the antioxidant enzymes that work together, being synthesized from different parts of the plant cell.

There are many studies reporting the As toxicity level and its effect on the germination capacity of subjected rice and wheat seeds (Shri et al., 2009; Abedin and Meharg, 2003; Mahdieh, Ghaderian and Karimi, 2013). Seeds with a different spiked concentration of arsenic (both As(III) and As(V)) were observed to have decreased germination rate as well as growth hindrance of the rice and wheat seedlings. Reports state that arsenite (As2O3) produces $O2^{\cdot-}$ and H_2O_2 in abundance and triggers lipid peroxidation in rice seedlings, followed by consequent anti-stress enzyme activity (Shri et al., 2009). Productions and activities of antioxidant enzymes like SOD, CAT, GR, GPX, APX, MDHAR and DHAR get fluctuated due to As(III) exposure that in turn induces the plant cell glutathione and it's derivative phytochelatins production. In the schematic Figure 1, the relation between arsenic toxicity in rice or wheat plant seedlings and bioaccumulation of arsenic, the biomass of plants root and shoot in seedlings, production of reactive oxygenic species and effect of germination rate has been shown. With the application of different concentration of arsenic in rice and wheat seedlings, bioaccumulation of arsenic differs with subjected rice and wheat species as well as their biomass ratio in root and shoot. Studies suggest that bioaccumulation of root arsenic is higher in rice plant root system particularly the As(III) in a hydroponic system. Not only the hydroponic culture, in field environments also, when waterlog condition prevails, the redox state of arsenic pentavalent gets changed to arsenic trivalent and that get their access to the plant root aquaporin channels to pass through from the surrounding water to the plant system, resulted in accumulation in the root zone. Seedling of both rice and wheat plants need water at the very beginning and that allows the soil arsenic to get accumulated in the plant. With the continuous uptake of arsenic, the stress on normal cellular activities in seedlings triggers to generate ROS and that in turn reduce the biomass of the subjected plants. Root and shoot with stressed response process try to cope up with their defense strategies that involve the nutritional deficiency to the whole plant body resulted in a decrease in the growth and biomass. On the other hand, during the seed soaking or the pre-processing of the seeds before plantation, studies have shown that arsenic spiking has a definite effect on germination capacity of rice and wheat plants irrespective of species differentiation. Study by Akhtar and Shoaib (2014) Germination percentage gets reduced from 80% in control without arsenic spiked soil to 25% in arsenic spiked soil with four different concentrations of arsenic and this was also correlated to the relative arsenic injury rate in seedlings of wheat from 50% to 75% in the least to highest arsenic spiked seedlings respectively. In some other study (Mahdieh, Ghaderian and Karimi, 2013), application of arsenic more than 10 mg/Kg in concentration started in the decrease of relative shoot height (RSH) afterward and that also affects the total chlorophyll content of the wheat plant leaves. With application gradient of arsenic up to 30 mg/Kg, relative chlorophyll content decreases significantly as the biomass of leaves declined. In case of rice cultivation, prolong and periodical application of water is needed and that allows to get concentrated with arsenite [As(III)] in rice plant's root, shoot, leaf, husk and grain along with some presence of arsenate [AS(V)]; but wheat cultivation does not need any such waterlog conditions to be maintained and dry conditions prevail most of the time. This allows the As(V) to be taken up by the wheat plants as in such dry soil, iron plaques surrounding the root system makes interaction with arsenate and increase the chances of its absorption on to the root surface from where the As(V) get transported to the plant system via the phosphate transporters. Although, both the inorganic form of arsenic becomes toxic to some the plant after a certain accumulation inside the plant cells or tissues, but exposure to As(III) imparts more toxic damages compare to the As(V) and that leads to the fact that generation of arsenic toxicity or other ROS elements are less in wheat compared to rice plants.

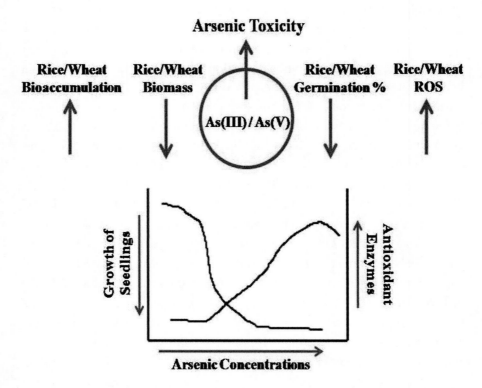

Figure 1. Schematic toxicologic effects of arsenic on rice and wheat plant seedlings. The schematic representation points out to the four major effect of arsenic toxicity on plant seedlings with negative effect indicated by down arrow, positive effect indicated by up arrow along with probable graphical hypothesis of anti-stress enzyme generation with increasing As concentration in plant cell matrix whereas growth of the subjected plant seedlings get decreased.

3. CELLULAR METABOLISM OF ARSENIC

3.1. Reduction of Arsenic in Cellular Matrix

Primarily, in the course of arsenic toxicity, plants develop some metabolic strategies that would help to sequester partial arsenic contamination from the plant tissues and that mainly acquired by two basic processes, reduction, and methylation of arsenic. Previous reports of arsenic accumulation in rice plant showed that amendment of arsenate to the surrounding media accounts a significant presence of arsenite in the root cell cytosol which clearly indicates the intracellular reduction of arsenate after its uptake and transport to the subsequent root cells to xylem parenchyma (Zhao et al., 2009). In yeast cells, a protein (Acr2p) from the tyrosine phosphatase superfamily was isolated and analysed in the process of arsenate reduction in yeast cells. Homologous to this protein Acr2p, in plants, a similar type of proteins have been discovered and analysed in the same course. The isolated protein from rice plant is *ACR2* that also works same as like *CDC25*-like protein (cell division cycle protein, a specific cell cycle dual-specificity phosphatases) along with two isoforms of rice protein *OsACR2* that functions like phosphatase enzyme (Dhankher et al., 2006). Some studies involved the expression of plant genes into microbes to check its applicability and justification in terms of arsenate reduction by engaging both wild type and mutant type of microbes like E. coli and yeast cells. In E. coli, expression of rice *OsACR2* was observed in the absence of *ArsC* gene or by knocking out the *ArsC* gene that is responsible for the arsenic resistance in microbes and that showed the reversion of arsenate reduction to arsenite, indicating the activity of *OsACR2* gene in the reduction process of As(V) to As(III) in rice plants (Ellis et al., 2006). In some other studies (Bleeker et al., 2006), purified protein *ACR2* from this selected *OsACR2;1* and *OsACR2;2* genes also was observed to work in conjugation with glutathione (GSH) and influence positively to mediate the GSH-based reduction of arsenate. Rice root cells mainly overexpress the *OsACR2;2* in association with *OsACR2;1* but in soot cell cytosol, mostly *OsACR2;1* is found. Some other studies reported that several genes are present in the rice plant cells that either up-regulate or down-regulate the conversion of arsenate as well as other toxic metal(loids) including subsequent gene regulation that results in the defense against arsenic toxicity after its heavy exposure (Tuli et al., 2010). Transcriptomic studies of rice seedlings also suggest that after one day from the exposure of arsenate, rice root cells triggers their cellular metabolism to up-regulate 81 different genes and 240 genes to down-regulate. The report also suggests that the plant abscisic acid (ABA), salicylic acid (SA), gibberellic acid (GA), flavonoids control the arsenic stress and subsequent toxicity in the course of arsenic metabolism (Chakrabarty et al., 2009).

3.2. Methylation of Organoarsenic Compounds

In a previous study (Nissen and Benson, 1982), researchers have shown that the nutrients depleted medium (lacking nitrogen and phosphorus) can be used to grow plants like tomato and that media was induced with radioactive arsenic species (As^{74}) to observe the translocation of arsenic through the root system. The study result found that the plant root cells significantly converted the As^{74} to some methylated species. But contrastingly, plants in nutrient enriched media were also found to consist of methylated arsenic species that was converted from that radioactive arsenate solution. Later on, it was found in several other studies that before translocation from the surrounding environment to the root cells or afterward to the upper plant body parts, organoarsenic compounds like MMAA, DMAA and trimethylarsine oxide (TMAO) get passed through via this methylation processes (Mihucz et al., 2005). In soil, microbes play a major role to convert these organoarsenic compounds into methylated forms and thus help indirectly in the transfer and bioavailability of such compounds. But in some hydroponic experiments, the presence of methylated organoarsenic compounds was found in the root cells as well as plant xylem sap, translocated further in the upper body parts, although quite lesser than the inorganic forms of arsenic being transported through the plant system. In rice, studies have shown that DMAA can be translocated through the plant xylem and phloem system and gets accumulated into the rice grains accounting negligible to approximately 90% of the total arsenic present in the seed, depending upon the cultivation process as well as the rice varieties (Meharg et al., 2008). In some report, experiments were carried out using both inorganic and organic arsenic species with different pH conditions maintained and that markedly influenced the uptake of these methylated compounds above pH 4. MMAA was taken up less above pH4.5 to 6.5 and above5.5 to 6.5 pH DMAA translocation was reduced (Li et al., 2009). In rice plants, the proper intracellular methylation pathway of these organoarsenic compounds, is yet to be discovered, although, analysis of whole rice plant genome suggests a possible gene presence that has a similar protein motif like UbiE/Coq5 protein family, which is present in the *arsM* gene from the microbial origin. The report has conducted with the involvement of rice root microarray experiment where they have found the *Os02g51030* gene with possible methyltransferase activity (Norton et al., 2008). The study by Li et al., (2009), suggests that rice plant root system can mediate the conversion from MMAA(V) to MMAA(III) by reduction via Challenger pathway, although there was no report of the mediator enzymes involvement or proper identification of those enzymes.

4. INDEMNIFICATION VIA DETOXIFICATION OF ARSENIC FROM PLANT CELLS

The toxicity of arsenic arises when arsenate [As(V)] interferes with phosphate metabolism and its phosphorylation along with ATP production whereas the arsenite [AS(III)] hinders the protein chemistry by binding to the sulphydryl groups (Hughes, 2002). These actions alter the in vivo interactions between cellular components and trigger the ROS components which in turn produce some of the anti-stress components including enzymes as well as non-enzymes molecules like glutathione and its derivative phytochelatins. Glutathione (GSH) is a tripeptidyl molecule involving glutamate-cysteine-glycine containing a thiol group also, which, being nucleophilic in nature performs as the electron acceptor or donor in diversified biochemical reactions. In the presence of toxic metals or metalloids, plant cells start to synthesize GSH which is a two step energy dependent reactions. Figure 2 depicts the glutathione and phytochelatins synthesis in plant root and leaf cell matrix as well as their translocation from the outside environment to the inside cellular matrix with passage through the xylem and phloem tissues.

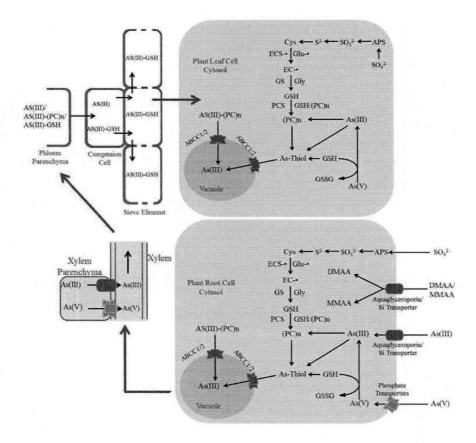

Figure 2. Schematic representation of GSH and PCs activity along with translocation of arsenic in plant systems. The network consists of four parts of arsenic translocation involving different transporter molecules from root to xylem to phloem to plant leaf cell as well as the synthesis pathway of glutathione in root and leaf cell matrices. Abbreviations are mentioned in full in respective text.

After the sulfate molecules get entry into the plant cells, in the course of GSH synthesis, L-glutamate and L-cysteine produces the γ-glutamylcysteine (EC) in presence of γ-glutamylcysteine synthetase (ECS). Next, the glutathione synthetase works on the previous compound to form glutathione (GSH) by adding a glycine to the C-terminal domain of the EC. GSH, in normal biological status, gets oxidized to glutathione disulfide (GSSG) which in turn gets reverted back to GSH due to the activity of glutathione reductase (GR) for maintaining the balance of the cellular function (Kao, 2015). The GR was first reported in wheat germ cells (Conn and Vennesland, 1951) and later was identified in rice plants also (Kaminaka et al., 1998). In the presence of any abiotic stress like arsenic toxicity, the development of ROS hampers the normal ratio of GSH/GSSG and that, on the other hand, affects the GR concentration also. In this course of stress, glutathione S-transferase (GST) controls the detoxification process involving the GSH electrophilic interactions. Apart from the GST, GPX enzyme is another important ROS scavenging enzyme molecule from the group of multiple isozymes family that reduces the hydrogen peroxides and lipid hydroperoxides in presence of GSH (Kao, 2015). Phytochelatins (PC) on the other hand, are metal binding peptide molecule with a general formula of (γ-Glu-Cys)n-Gly, where n is the repetitive number of this unit from 2 to 20, having a strong affinity to bind with the arsenic and cadmium species and subsequent detoxification. PC is a derivative of GSH and the enzyme that synthesizes PC from GSH is called as phytochelatin synthase, performs a transpeptidation reaction by removing a Glu-Cys moiety from the donor to the acceptor molecule. Studies have shown the importance of PC in terms of arsenic tolerance and subsequent detoxification process (Cobbett, 2000). Inhibiting the PC synthesis by adding L-buthionine-sulphoxime (BSO), an inhibitor of γ-glutamylcysteine synthetase makes the subjected plant hypersensitive to the presence of arsenic. The conjugation of PC-As has been isolated from the plant tissues by using high-performance liquid chromatography with inductively coupled plasma mass spectrometry (HPLC-ICP-MS) and molecule-specific electrospray-ionization mass spectrometry (ES-MS), identifying various species of PC-As complexes like As(III)-(PC2)$_2$, As(III)-PC3, MMA-PC2 and some GSH-As bound PCs (Raab et al., 2005). Some of these complexes are stored in the plant cell vacuoles and transported out of the plant cell via transporter molecules like ABCC1/2 that is a member of the ATP-binding cassette (ABC) superfamily (Bleeeker et al., 2006). Figure 2 has shown the overview of arsenic translocation from root cell cytosol to its final destination of plant leaf cell cytosol via xylem and phloem pathway. In both root and leaf cell matrix, GSH and PCs play their role to chelate down or detoxify the arsenic, present in the cellular matrix.

SUMMARY

Arsenic toxicity is an expanding concern from every aspect including sustainable farming, to consumer's health risk parameters, toxicology profile of cereal plants like rice and wheat, subsequent retardation of arsenic contamination in food products and much more. Several studies have investigated the cause and effect of arsenic toxicity in rice and wheat plants in all the stages of its growth with bio-accumulation and cellular damage assessment. Transportation of arsenic depending on its speciation translocates via aquaglyceroporins silicon and phosphate transporters. Due to the increased abiotic stress on the rice and wheat plants, various reactive oxygenic species are been triggered on which in return causes the generation of several anti-stress enzymes and some non-enzymatic molecules including glutathione and phytochelatins that play crucially in the detoxification and tolerance process of arsenic inside the cellular matrix. Reduction of arsenate to arsenite with complementary methylation process helps to alleviate the toxicity in plant cells to some extent. The arsenic toxicogenesis profile is utterly crucial for proper management with required development of the arsenic remediation in rice and wheat like cereal crops.

ACKNOWLEDGMENTS

Authors are thankful to IISER K for providing infrastructure and basic research facilities. AM is thankful to Ministry of Earth Sciences (MoES/P.O. (Geosci)/56/2015) for providing fund in the form of JRF and SB is thankful to SERB-DST, Government of India for providing fund in the form of Ramanujan Research Grant (SR/S2/RJN-09/2011) to carry all research works.

CONFLICT OF INTEREST

The authors (Arnab Majumder and Sutapa Bose) declare that is no conflict of interest regarding the publication of this manuscript.

REFERENCES

Abedin, M. J. and Meharg, A. A., 2002. Relative toxicity of arsenite and arsenate on germination and early seedling growth of rice (Oryza sativa L.). *Plant and soil*, 243(1), pp. 57-66.

Akhtar, S. and Shoaib, A., 2014. Toxic effect of arsenate on germination, early growth and bioaccumulation in wheat (Triticum aestivum L.). *Pakistan Journal of Agricultural Sciences*, 51(2), pp. 399-404.

Barla, A., Shrivastava, A., Majumdar, A., Upadhyay, M. K. and Bose, S., 2017. Heavy metal dispersion in water saturated and water unsaturated soil of Bengal delta region, India. *Chemosphere*, 168, pp. 807-816.

Bhattacharjee, H. and Rosen, B. P., 2007. Arsenic metabolism in prokaryotic and eukaryotic microbes. In *Molecular microbiology of heavy metals* (pp. 371-406). Springer Berlin Heidelberg.

Bleeker, P. M., Hakvoort, H. W., Bliek, M., Souer, E. and Schat, H., 2006. Enhanced arsenate reduction by a CDC25-like tyrosine phosphatase explains increased phytochelatin accumulation in arsenate-tolerant Holcus lanatus. *The Plant Journal*, 45(6), pp. 917-929.

Bucher, M., 2007. Functional biology of plant phosphate uptake at root and mycorrhiza interfaces. *New Phytologist*, 173(1), pp. 11-26.

Chakrabarty, D., Trivedi, P. K., Misra, P., Tiwari, M., Shri, M., Shukla, D., Kumar, S., Rai, A., Pandey, A., Nigam, D. and Tripathi, R. D., 2009. Comparative transcriptome analysis of arsenate and arsenite stresses in rice seedlings. *Chemosphere*, 74(5), pp. 688-702.

Cobbett, C. S., 2000. Phytochelatins and their roles in heavy metal detoxification. *Plant physiology*, 123(3), pp. 825-832.

Conn, E. E. and Vennesland, B., 1951. Glutathione reductase of wheat germ. *Journal of Biological Chemistry*, 192(1), pp. 17-28.

Dhankher, O. P., Rosen, B. P., McKinney, E. C. and Meagher, R. B., 2006. Hyperaccumulation of arsenic in the shoots of Arabidopsis silenced for arsenate reductase (ACR2). *Proceedings of the National Academy of Sciences*, 103(14), pp.5413-5418.

Ellis, D. R., Gumaelius, L., Indriolo, E., Pickering, I. J., Banks, J. A. and Salt, D. E., 2006. A novel arsenate reductase from the arsenic hyperaccumulating fern Pteris vittata. *Plant physiology*, 141(4), pp. 1544-1554.

Hirayama, T. and Shinozaki, K., 2010. Research on plant abiotic stress responses in the post-genome era: Past, present and future. *The Plant Journal,* 61(6), pp.1041-1052.

Huang, J. H. and Matzner, E., 2006. Dynamics of organic and inorganic arsenic in the solution phase of an acidic fen in Germany. *Geochimica et Cosmochimica Acta*, 70(8), pp. 2023-2033.

Hughes, M. F., 2002. Arsenic toxicity and potential mechanisms of action *Toxicol Lett* 133: 1–16.

Kaminaka, H., Morita, S., Nakajima, M., Masumura, T. and Tanaka, K., 1998. Gene cloning and expression of cytosolic glutathione reductase in rice (Oryza sativa L.). *Plant and cell physiology*, 39(12), pp. 1269-1280.

Kao, C. H., 2015. Role of glutathione in abiotic stress tolerance of rice plants. *J. Taiwan Agric. Res.* 64(3), pp. 167-176.

Li, R. Y., Ago, Y., Liu, W. J., Mitani, N., Feldmann, J., McGrath, S. P., Ma, J. F. and Zhao, F. J., 2009. The rice aquaporin Lsi1 mediates uptake of methylated arsenic species. *Plant Physiology*, 150(4), pp. 2071-2080.

Liu, W. J., Zhu, Y. G., Hu, Y., Williams, P. N., Gault, A. G., Meharg, A. A., Charnock, J. M. and Smith, F. A., 2006. Arsenic sequestration in iron plaque, its accumulation and speciation in mature rice plants (Oryza sativa L.). *Environmental science & technology*, 40(18), pp. 5730-5736.

Ma, J. F., Tamai, K., Yamaji, N., Mitani, N., Konishi, S., Katsuhara, M., Ishiguro, M., Murata, Y. and Yano, M., 2006. A silicon transporter in rice. *Nature*, 440(7084), pp. 688-691.

Mahdieh, S., Ghaderian, S. M. and Karimi, N., 2013. Effect of arsenic on germination, photosynthesis and growth parameters of two winter wheat varieties in Iran. *Journal of plant nutrition*, 36(4), pp. 651-664.

Meharg, A. A., 2004. Arsenic in rice–understanding a new disaster for South-East Asia. *Trends in plant science*, 9(9), pp. 415-417.

Meharg, A. A., Lombi, E., Williams, P. N., Scheckel, K. G., Feldmann, J., Raab, A., Zhu, Y. and Islam, R., 2008. Speciation and localization of arsenic in white and brown rice grains. *Environmental Science & Technology*, 42(4), pp. 1051-1057.

Mihucz, V. G., Tatár, E., Virág, I., Cseh, E., Fodor, F. and Záray, G., 2005. Arsenic speciation in xylem sap of cucumber (Cucumis sativus L.). *Analytical and bioanalytical chemistry*, 383(3), pp. 461-466.

Nissen, P. and Benson, A. A., 1982. Arsenic metabolism in freshwater and terrestrial plants. *Physiologia Plantarum*, 54(4), pp. 446-450.

Norton, G. J., Lou-Hing, D. E., Meharg, A. A. and Price, A. H., 2008. Rice–arsenate interactions in hydroponics: whole genome transcriptional analysis. *Journal of Experimental Botany*, 59(8), pp. 2267-2276.

Poynton, C. Y., Huang, J. W., Blaylock, M. J., Kochian, L. V. and Elless, M. P., 2004. Mechanisms of arsenic hyperaccumulation in Pteris species: root As influx and translocation. *Planta*, 219(6), pp. 1080-1088.

Raab, A., Schat, H., Meharg, A. A. and Feldmann, J., 2005. Uptake, translocation and transformation of arsenate and arsenite in sunflower (Helianthus annuus): formation of arsenic–phytochelatin complexes during exposure to high arsenic concentrations. *New phytologist,* 168(3), pp. 551-558.

Sahoo, P. K. and Mukherjee, A., 2014. Arsenic fate and transport in the groundwater-soil-plant system: an understanding of suitable rice paddy cultivation in arsenic enriched areas. In *Recent Trends in Modelling of Environmental Contaminants* (pp. 21-44). Springer India.

Shin, H., Shin, H. S., Dewbre, G. R. and Harrison, M. J., 2004. Phosphate transport in Arabidopsis: Pht1; 1 and Pht1; 4 play a major role in phosphate acquisition from both low-and high-phosphate environments. *The Plant Journal*, 39(4), pp. 629-642.

Shri, M., Kumar, S., Chakrabarty, D., Trivedi, P. K., Mallick, S., Misra, P., Shukla, D., Mishra, S., Srivastava, S., Tripathi, R. D. and Tuli, R., 2009. Effect of arsenic on growth, oxidative stress, and antioxidant system in rice seedlings. *Ecotoxicology and environmental safety,* 72(4), pp. 1102-1110.

Shrivastava, A., Barla, A., Yadav, H. and Bose, S., 2014. Arsenic contamination in shallow groundwater and agricultural soil of Chakdaha block, West Bengal, India. *Frontiers in Environmental Science*, 2, p. 50.

Tuli, R., Chakrabarty, D., Trivedi, P. K. and Tripathi, R. D., 2010. Recent advances in arsenic accumulation and metabolism in rice. *Molecular Breeding,* 26(2), pp. 307-323.

You, J. and Chan, Z., 2015. ROS regulation during abiotic stress responses in crop plants. *Frontiers in plant science*, 6, p. 1092.

Zhao, F. J., Ma, J. F., Meharg, A. A. and McGrath, S. P., 2009. Arsenic uptake and metabolism in plants. *New Phytologist*, 181(4), pp. 777-794.

Zhao, F. J., McGrath, S. P. and Meharg, A. A., 2010. Arsenic as a food chain contaminant: mechanisms of plant uptake and metabolism and mitigation strategies. *Annual review of plant biology*, 61, pp. 535-559.

INDEX

X

Y

Z